The When, Where, and How of Morphological Processing in the Human Brain

Fei Gao

The When, Where, and How of Morphological Processing in the Human Brain

Evidence from Mandarin Chinese

Springer

Fei Gao
Fudan University
Shanghai, China

ISBN 978-981-95-5980-0 ISBN 978-981-95-5981-7 (eBook)
https://doi.org/10.1007/978-981-95-5981-7

This Springer imprint is published by the registered company Springer Nature Singapore Pte Ltd.
The registered company address is: 152 Beach Road, #21-01/04 Gateway East, Singapore 189721, Singapore

To my departed, mysterious father
Gao Daqin
高大勤 *(1961–2008)*

Acknowledgments

Motivated by the lexical production errors observed among the second language learners of Chinese, I began to focus on issues of lexical processing and learning in 2016. Over the following decade, my research has centered on Chinese morphological processing and vocabulary learning. Most existing reading research has concentrated on Indo-European languages, especially English. Yet, given that Chinese is the most widely spoken language in the world, evidence from Chinese is essential for testing and refining general theories of reading science, especially since Chinese differs substantially from Indo-European languages in its typological characteristics. I also realized that, compared with semantics and syntax, morphology has been relatively understudied, particularly from the perspectives of cognitive processing and neural representation. The psychological reality of Chinese morphological processing, as well as its similarities and differences with Indo-European languages, remains insufficiently understood. Therefore, in addition to employing traditional corpus-based and behavioral methods, I have also emphasized tackling these scientific questions from the perspective of neural representation.

This book is devoted to Chinese morphology and morphological processing. By combining the unique linguistic perspective of Chinese with multimodal neuroimaging techniques, I hope to provide novel insights into cross-linguistic reading and learning research. The book is based on my PhD thesis (*The When, Where, and How of Chinese Morphological Processing in the Human Brain: Evidence from EEG-fNIRS*, 2022, University of Macau) and partially on my master's thesis (*The Representation of Chinese Second Language Learners' Lexical Coinages—Behavioral and Electrophysiological Evidence*, 2018, Beijing Language and Culture University). Some of the data from these theses have appeared in journal publications. For the purposes of this book, I have reanalyzed and/or reinterpreted the thesis data in order to highlight a cross-linguistic comparative perspective and to frame the findings within an overarching theoretical account, i.e., the dual-route model. This framework emphasizes the dynamic interplay of linguistic features and individual differences in shaping processing and learning strategies. Another highlight of this book is the adoption of an innovative integration of Electroencephalography (EEG) and functional Near-Infrared Spectroscopy (fNIRS), which allows for a fine-grained investigation of both the temporal dynamics and neural localization of Chinese morphological processing, thereby advancing our theoretical understanding.

The first four chapters review the research progress in Chinese reading and methodological developments, with a particular emphasis on Chinese morphology and morphological processing. Chapters 5–8 present four experimental studies that address the "when, where, and how" of Chinese morphological processing in the human brain across different aspects and scenarios. These studies are designed in a progressive sequence, culminating in the final chapter with implications and applications. By integrating perspectives from linguistics, psycholinguistics, and neuroscience, this book aims to be of value to researchers across these fields.

This book would not have been possible without the guidance and support of many people. First and foremost, I am deeply grateful to Professor Jianqin Wang, who first introduced me to psycholinguistics and provided crucial guidance on methodology. His integrity and curiosity have left a lasting impact on me, and his encouragement since 2015 has been indispensable. I also owe profound thanks to Professor Zhen Yuan, who welcomed me at a difficult moment and introduced me to the broad world of bio-imaging, thereby expanding my research horizons. My heartfelt gratitude also goes to Professor Cecilia Zhao, whose professionalism and passion for research have inspired me greatly, and whose assistance and support in writing my first English article was invaluable. Her supervision style has deeply influenced my own approach to mentoring students. I feel fortunate to have had the guidance of such professors, who have largely shaped my academic voice.

I would also like to thank Zeping Liu, Jingting Ye, Qian Yang, Jun Ouyang, Sichang Gao, Henry Wang, Jimmy Zhong, and Tian Zhang, for proofreading the chapters of this book. My deepest appreciation goes to my mother Mingqing Lyu and sister Tiantian Gao for their unconditional love and support throughout my personal growth and academic journey. I am grateful to my friends, Meng-Yun Wang, Miriam Akioma, Shiyang Xu, Owen He, Vincent Zeng, and Lin Hua for their company and assistance in research, as well as to my colleague Alex Smith for his moral support—we wrote our monographs side by side. Finally, I thank everyone I have encountered in life and this beautiful world that continues to inspire me.

Abbreviations

ACC	accuracy
ANOVA	analyses of variance
AoA	age of acquisition
BA	Brodmann area
BCI	brain–computer interface
DLPFC	dorsolateral prefrontal cortex
ECoG	electrocorticography
EEG	electroencephalography
ELAN	early left-anterior negativity
EmoPro	Emotional Prototypicality
ERP	event-related potential
fMRI	functional Magnetic Resonance Imaging
fNIRS	functional Near-Infrared Spectroscopy
FUS	focused ultrasound
GLM	general linear model
HbO	oxygenated hemoglobin
HbR	deoxygenated hemoglobin
HbT	total hemoglobin
HRF	hemodynamic response function
ICA	independent component analysis
IIC	Inter/Intra Connection
iEEG	intracranial electroencephalography
IFG	inferior frontal gyrus
L1	first language
L2	second language
L3	third language
LAN	left anterior negativity
LLM	large language model
MEG	magnetoencephalography
MFG	middle frontal gyrus
MNI	Montreal Neurological Institute

MTG	middle temporal gyrus
MVPA	multivariate pattern analysis
NLP	natural language processing
PET	positron emission tomography
PTG	posterior temporal gyrus
RD	reading disability
ROI	region of interest
RSA	representational similarity analysis
RT	reaction time
SER	sensory experience rating
sNOM	subjective number of meanings
SOA	stimulus onset asynchrony
SPM	statistical parametric modeling
STG	superior temporal gyrus
tDCS	transcranial direct current stimulation
TMS	transcranial magnetic stimulation
tPBM	transcranial photobiomodulation
VF	half-field
VWFA	visual word form area

Contents

Chapter 1
Introduction

1.1 Research Background and Motivation

Morphology, the study of the internal structure of words, has long been regarded as a cornerstone of linguistic theory, reading development, and second language (L2) acquisition. Morphological processing refers to the way in which speakers and readers perceive, access, and manipulate morphemes during comprehension and production. It constitutes a fundamental mechanism for building lexical representations and for supporting efficient word recognition. Over the past decades, research in psycholinguistics and neurolinguistics has shown that morphological knowledge not only facilitates vocabulary growth and literacy development but also reflects essential principles of how language is organized and processed in the human mind and brain.

Chinese, spoken by the largest population in the world, occupies a unique position in this field. More than one quarter of the global population relies on Chinese characters for reading (Tseng 1982). Nevertheless, the scientific study of reading has been dominated by English, with insufficient attention to other languages (Share 2008). The investigation of Chinese, with its striking typological contrast to alphabetic systems, provides a critical testing ground. Findings from Chinese can support or challenge conclusions drawn from alphabetic languages and thereby contribute to the development of a universal science of reading (Perfetti et al. 2013).

As a morpho-syllabic system (DeFrancis 1989), Chinese morphology presents salient differences from alphabetic languages. Yet it remains unclear how the human brain encodes and processes different aspects and types of morphological information during Chinese word reading. This book therefore begins by examining the salience of Chinese morphology and its neural bases in lexical representation and processing. More specifically, we focus on three interrelated questions: *when* morphological information is accessed during word recognition, *where* in the brain such processing takes place, and *how* different types of morphological structures are

F. Gao, *The When, Where, and How of Morphological Processing in the Human Brain*, https://doi.org/10.1007/978-981-95-5981-7_1

represented and integrated. These three dimensions together frame the investigation of Chinese morphological processing and provide a comprehensive approach to linking linguistic theory with cognitive and neural evidence.

At the same time, Chinese morphology plays a crucial role in linguistic and psycholinguistic theories as well as in the organization of the mental lexicon of native speakers. However, relatively little is known about how Chinese morphological processing unfolds across different learning contexts. For instance, the mechanisms underlying morphological processing in L2 learners of Chinese remain largely unexplored, despite their importance for understanding how non-native readers acquire and utilize morphological cues in a morpho-syllabic system. In addition, questions remain regarding whether and how Chinese–English bilinguals can transfer morphological processing resources from Chinese to English, given the profound structural differences between the two languages. This book therefore addresses Chinese morphological processing in multiple scenarios, including native speakers, L2 learners of Chinese, and adult Chinese–English bilinguals. By adopting this multi-faceted perspective, the book aims to provide a comprehensive account of both language-specific and language-general principles of morphological processing in the human brain.

1.2 Research Questions

Despite the increasing recognition of morphology as a crucial component of the Chinese writing system, several research gaps remain regarding the neurocognitive mechanisms of Chinese morphological processing. First, it is still debated whether the fundamental representational unit in the mental lexicon of native Chinese speakers is whole-word based or decomposed, and whether L2 learners of Chinese develop similar representational patterns. Second, the temporal dynamics of Chinese morphological processing have not been clearly established. Specifically, it is uncertain whether sensitivity to morphological structure reflects automatic processes or conscious and controlled evaluation. Third, the neural localization of Chinese morphological processing remains poorly understood, particularly with respect to how morphological constraints can be dissociated from semantic processing in the cortex, and whether different morphological structures leave distinct spatial signatures in the brain. Finally, very little is known about the neural correlates of morphological processing in bilingual adults, and it remains unclear whether Chinese and English rely on shared or distinct brain networks when processing morphological information, given the significant language distance across the two languages.

In light of these research gaps, this book addresses the following interconnected research questions:

1. What are the fundamental morphological units in the mental lexicon of Chinese first language (L1) speakers, and do L2 learners of Chinese develop comparable representational patterns?
2. What are the temporal signatures and spatial activations that characterize the encoding of morphological constraints in Chinese compound words, and how do these differ from semantic processing in the brain?
3. What are the temporal and spatial specifications of the word structure effect in Chinese word reading, and do different morphological structures engage common or distinct neural mechanisms?
4. To what extent does morphological processing in Chinese recruit specific cognitive and neural resources, and can these resources be extended to languages with greater typological distance, such as English, thereby informing a broader theory of language processing?

To address these questions, this book combines corpus-based analyses, behavioral measures, and multiple neuroimaging techniques. Together, these research questions align with the overarching framework of this book, which seeks to reveal *when* morphological information is accessed in Chinese word reading, *where* in the brain such processes are localized, and *how* different morphological structures are encoded and integrated.

1.3 Significance of the Book

To the best of our knowledge, this book represents the first systematic set of studies that relate the dynamics of critical brain events to morphological information, including units, constraints, structures, and bilingual representations, during the processing of single Chinese words. In particular, it provides the first neuroimaging evidence for the neurocognitive underpinnings of Chinese derivational word reading, a domain that has received little attention in the existing literature.

The findings of this book make important theoretical contributions. By investigating Chinese morphological processing, the research tests and extends several influential models developed primarily on Indo-European languages, such as the interactive activation model (Taft 1994), the dual-route account (Clahsen 1999), and the transfer facilitation model (Koda 2005). Through the lens of Chinese morphology, these models are evaluated in a typologically distinct language, thereby broadening their scope and refining their explanatory power.

Beyond theoretical advances, this book also carries practical implications for language learning and pedagogy. The studies on L2 learners of Chinese and Chinese–English bilinguals provide new insights into how morphological awareness and processing skills develop across languages. Such insights can inform evidence-based approaches to teaching Chinese as a second language, as well as bilingual literacy instruction more broadly, by highlighting the role of morphological knowledge in reading acquisition and cross-linguistic transfer.

From a methodological perspective, this book demonstrates the utility of a novel multimodal approach that combines electrophysiological and optical neuroimaging methods within a single study. The simultaneous use of the two imaging modalities allows for the integration of temporal and spatial dimensions of word reading, thereby offering a practical and replicable framework for multimodal data fusion in reading science. In addition, we advocate a three-stage research framework of corpus-driven observation, experimental investigation, and implications/applications. It situates multimodal neurocognitive evidence within a broader cycle of discovery and application in morphological processing and reading research.

In sum, the book not only deepens our understanding of the idiosyncrasies of the Chinese morphological system but also contributes to the construction of a general neural architecture of word reading across languages, bridging linguistic theory, cognitive neuroscience, and educational practice.

1.4 Structure of the Book

This book is organized into nine chapters.

Chapter 1 introduces the overall research background and motivation. This chapter also identifies critical research gaps, formulates the guiding research questions, and outlines the significance of the book in terms of its theoretical and methodological contributions as well as real-world implications.

Chapter 2 provides a comprehensive review of the past two decades of cognitive and neural research on Chinese reading. This chapter sets the stage by situating morphology and morphological processing within the broader landscape of reading studies.

Chapter 3 offers a systematic introduction to morphology and morphological processing. Beginning with Indo-European languages and extending to Chinese, it covers definitions, classifications, and theoretical frameworks. It then discusses key research issues, current trends, and critical variables in the study of Chinese morphological processing.

Chapter 4 presents methodological approaches to the study of Chinese reading, with particular emphasis on morphological processing. It introduces the corpus-based approach, behavioral paradigms, neuroimaging techniques, and statistical models, which are employed in the empirical studies of this book.

Chapter 5 employs behavioral experiments to investigate the representational units in the mental lexicon of Chinese L1 speakers and L2 learners, with particular reference to the dual-route model.

Chapter 6 applies multimodal neuroimaging to examine the neural encoding of morphological constraints in Chinese. By integrating temporal and spatial perspectives, this chapter explores whether morphological and semantic processing can be dissociated at the neural level.

Chapter 7 extends the inquiry to differing morphological structures, testing the psychological and neural reality of compounding and derivation differentiation in

Chinese. Using a morphological priming paradigm, it compares the processing strategies associated with these two structures.

Chapter 8 broadens the scope to a bilingual context, investigating whether Chinese–English bilinguals can transfer the cognitive and neural resources underlying Chinese morphological processing to their L2 English.

Chapter 9 concludes the book by summarizing the main findings, highlighting contributions to existing theories and pedagogical practices, and outlining directions for future research.

References

Clahsen, H. (1999). Lexical entries and rules of language: A multidisciplinary study of German inflection. *Behavioral and brain sciences*, *22*(6), 991-1013.

DeFrancis, J. (1989). *Visible speech: The diverse oneness of writing systems*. University of Hawaii Press.

Koda, K. (2005). *Insights into second language reading: A cross-linguistic approach*. Cambridge University Press.

Perfetti, C., Cao, F., & Booth, J. (2013). Specialization and Universals in the Development of Reading Skill: How Chinese Research Informs a Universal Science of Reading. *Sci Stud Read*, *17*(1), 5-21. https://doi.org/10.1080/10888438.2012.689786

Share, D. L. (2008). On the Anglocentricities of current reading research and practice: the perils of overreliance on an "outlier" orthography. *Psychol Bull*, *134*(4), 584-615. https://doi.org/10.1037/0033-2909.134.4.584

Taft, M. (1994). Interactive-activation as a framework for understanding morphological processing. *Language and Cognitive Processes*, *9*(3), 271-294.

Tseng, O. J.-L. (1982). Relationship of word combination in reading process and cognitive abilities: Status of Chinese reading researches in the contemporary cognitive science. In H. S.-R. Kao & C.-M. Cheng (Eds.), *Psychology and Research of the Chinese Language* (pp. 85-101). Crane.

Chapter 2
Cognitive and Neural Perspectives on Chinese Reading

2.1 From Reading Research to Chinese Word Reading

Reading is a fundamental cognitive activity that enables humans to acquire knowledge, communicate ideas, and participate in cultural and social life. As a field of inquiry, reading research provides the broader foundation for investigating morphological processing, since it examines how written symbols are mapped onto linguistic representations and how these processes are supported by cognitive and neural mechanisms. In the case of Chinese, such research has yielded important insights into the distinctive cognitive and neural features of a morpho-syllabic writing system, thereby laying the groundwork for understanding Chinese morphological processing.

Over the past century, reading studies have developed into a multidisciplinary field that integrates linguistics, psychology, education, and neuroscience. Early investigations in the late nineteenth and early twentieth centuries focused on pedagogical and practical issues, such as literacy instruction (Huey 1908; Thorndike 1917). From the mid-twentieth century onward, experimental psychology provided rigorous methods to examine word recognition, comprehension, and learning (Gibson and Levin 1975). With the rise of cognitive psychology in the 1960s and 1970s, attention shifted toward the mental representations and processes underlying reading (LaBerge and Samuels 1974). Since the 1980s, technological advances, especially in neuroimaging, have made it possible to investigate the neural bases of reading and reading disorders such as dyslexia (Perfetti 1985; Pugh et al. 2000). Thus, reading research has evolved from pedagogical beginnings to an interdisciplinary enterprise that combines behavioral experiments with neurocognitive methods, making it central to understanding both typical and atypical language processing.

Against this broad background, Chinese reading research could date back to the 1920s, which includes word and character recognition, comprehension, learning,

F. Gao, *The When, Where, and How of Morphological Processing in the Human Brain*, https://doi.org/10.1007/978-981-95-5981-7_2

assessment, among others. In terms of research issues and methodologies, Tan and Li (2016) classified the psycholinguistic and neurolinguistic studies of Chinese reading into the following four phases: (1) early explorations into the characteristics of Chinese script from pedagogical and practical perspectives (1921–1948); (2) the empirical investigations about the cognitive processing of Chinese language drawing on conventional methods of experimental psychology (1950–1981); (3) the emergence of inter-disciplinary, cross-cultural, and integrated studies on the multi-layer processing of Chinese script (1982–1999), especially accumulating evidence from neurophysiology, brain sciences, and computer sciences; (4) extensive utilizations of neuroimaging methods in Chinese reading since 2000 marks a new era of the filed, making it possible to touch upon the neural underpinnings of Chinese language and link language functions to their corresponding anatomical structures. Importantly, as highlighted by Yang et al. (2022), Chinese neurolinguistics has increasingly focused on language-specific cognitive mechanisms rather than applying models derived from alphabetic languages directly to Chinese. They emphasized that the cognitive and neural basis of Chinese reading must be understood in its own terms, considering the script's unique morphological, orthographic, and semantic characteristics.

Among the existing studies and various linguistic components, words have always been recognized as the driving force and linking center in language comprehension and production systems (Levelt 1989; Perfetti and Stafura 2013). In particular, word print is one of the most remarkable achievements of the human individuals (Cain and Parrila 2013). However, the neurocognitive mechanisms underlying Chinese word reading are still under exploration and its related brain evidence is inadequate, especially when the linguistic distance between Chinese and western languages (e.g., English) is addressed. Chinese characters originated as pictographs, which means that many of them historically depicted concrete objects and continue to preserve a close link between form and meaning (DeFrancis 1989). Unlike alphabetic languages, moreover, Chinese relies minimally on inflection, encodes morphemic meaning at the character level, and is dominated by compounding, with two-character words being the most common form (Packard 2000; Taylor and Taylor 2014). As a result, Chinese word reading involves the mapping between orthography, syllabic phonology, and morphemic semantics, leading to distinctive processing routes such as reliance on semantic–phonetic compound characters and sensitivity to morpheme boundaries (Perfetti et al. 2005; Tan et al. 2005b). Importantly, the issue of how these mappings are enabled and constrained by the human brain is an integral part of reading science. It merits a systematic review in the new millennium and a research agenda for the future directions.

With this retrospective review, this chapter aims to survey and synthesize the major theories, recent advances, and current issues in Chinese word reading over the past two decades, with a particular focus on cognitive and neurolinguistic perspectives. We concentrated on electrophysiological and neuroimaging studies of single characters and words, thereby covering the core types of Chinese word forms. By narrowing the scope to electrophysiological and imaging evidence, this chapter seeks to provide a precise account of the brain mechanisms underlying Chinese word reading and to highlight the unique contributions of psycholinguistic and neurolinguistic approaches to this field.

2.2 Brain Differences in Chinese and Alphabetic Word Reading

Reading science has been advanced by comparative studies that reveal both script-universal and script-specific features across different writing systems (Anderson and Chen 2013). From a language typology lens, Chinese demonstrates the maximum contrast with alphabetic systems among the major writing systems of the world. First, at the script level, Chinese is a logographic writing system, which is distinct from the linear layout of the most alphabetic scripts. Each character corresponds to a syllable and a morpheme, instead of a phoneme. Therefore, the concept of "word/词 (/ci2/)" in Chinese script is more flexible and ambiguous than that in alphabetic systems (Chung et al. 2010). Morphology provides another striking contrast. Chinese lacks inflections or derivations; instead, most words are formed by compounding. Single-character words account for less than 10% of all Chinese words, while two-character (i.e., disyllabic) compound words take up more than 70% and thus serves as the most popular materials in experimental studies of Chinese language processing (Tsang et al. 2018). Given these linguistic characteristics and their distance from alphabetic systems, Chinese word reading provides a unique testing ground for understanding how the brain accommodates diverse mapping principles between orthography, phonology, and semantics. Psycholinguistic and neurolinguistic research has therefore devoted substantial efforts to uncovering the cognitive and neural mechanisms specific to Chinese word reading, highlighting both shared cognitive principles and language-specific adaptations.

Brain activations during word reading from Chinese and alphabetic backgrounds have shown both shared and distinct patterns. The universal account of brain regions responsible for reading (e.g., Bolger et al. 2005; Nakamura et al. 2012; Zhang et al. 2019) believed that there is a universal brain network for recognizing both word shapes (i.e., "*reading by eye*") and handwriting gestures (i.e., "*reading by hand*") during visual word reading. Rueckl et al. (2015) further proposed and tested a framework where a universal hallmark of proficient reading consists of the convergence of speech/orthography processing systems onto a common brain network, regardless of how spoken words and meanings are represented in a given writing system (e.g., Spanish, English, Hebrew, and Chinese). Studies of alphabetic languages lend support to this universal account, showing that there are three common brain regions involved in word reading, where phonological processing is believed to be associated with temporo-parietal regions (e.g., superior temporal gyrus, STG), orthographic processing with left occipito-temporal visual word form area (VWFA)/fusiform gyrus, and semantic operation with inferior frontal gyrus (IFG) and middle temporal gyrus (MTG) (e.g., Cao 2015; Chen et al. 2009; Jobard et al. 2003; Turkeltaub et al. 2002).

On the other hand, even though there are overlapping areas involved in word reading between Chinese and alphabetic languages (e.g., left IFG for semantic processing of Chinese characters, Chou et al. 2009), Chinese word reading has also been proposed to rely on regions outside the aforementioned classical network for alphabetic reading. For example, while alphabetic reading shows a strong left-lateralized effect in the left frontal gyrus (Price 2000; Turkeltaub et al. 2002), Chinese activates corresponding areas in both hemispheres (Bolger et al. 2005). This bilateral involvement is thought to reflect the processing of stroke-defined components in the left hemisphere, complemented by the right hemisphere's role in spatial analysis of the character's two-dimensional configuration (Perfetti et al. 2013).

Meanwhile, a key distinction between alphabetic and Chinese reading lies in their phonological processing routes. In alphabetic systems, phonological access typically relies on assembled phonology, whereby spoken forms are generated through the sequential combination of smaller phonemic units (Coltheart et al. 1993). By contrast, Chinese word reading depends more strongly on addressed phonology, a bottom-up process in which phonological information is accessed after retrieving the visuo-orthographic representation of the lexical candidate. Neuroimaging evidence (Tan et al. 2005a) supports this distinction: the left middle frontal gyrus (MFG) has been identified as the core region for the addressed phonology in Chinese, whereas the left temporoparietal cortex is primarily responsible for the assembled phonology in alphabetic languages. To reconcile these cross-linguistic patterns, a covariant learning hypothesis has been proposed, suggesting that different cortical systems contribute to phonological processing depending on the specific demands of the writing system.

Moreover, Tan and his colleagues (e.g., Tan et al. 2000, 2001) have worked on a set of functional magnetic resonance imaging (fMRI) studies, which suggest that the left MFG (Brodmann area-BA9) might be the crucial and unique area for Chinese reading. This region was then tested by following studies (Chen et al. 2008; Li et al. 2006). Drawing on fNIRS technology, for instance, Chen et al. (2008) revealed comparable patterns and suggested an important role of the left MFG for Chinese word reading and a special role of the left STG (BA22) and supramarginal gyrus (BA40) for English readers. These findings indicate that Chinese word reading is characterized by a neural architecture that both overlaps with and diverges from alphabetic reading systems, reflecting the unique demands of mapping complex visuo-orthographic forms onto morphemic and syllabic representations.

Nevertheless, a recent meta-analytic study (Zhao et al. 2017) suggested that previously identified script-specific brain regions (e.g., MFG from Chinese reading, Tan et al. 2005a, 2005b; posterior superior temporal gyrus, pSTG for alphabetic reading, Paulesu et al. 2000) could be task dependent rather than stimulus selective. By summarizing the brain regions in previous fMRI studies from stimulus contrasts (SC, e.g., Chinese vs. Korean words) and task contrasts (TC), Zhao and colleagues revealed a shared network for two types of comparisons, while the SC > TC and

TC > SC comparisons demonstrated distinct activation systems. In particular, the TC > SC comparison showed significant involvement in the IFG, MFG, pSTG, and pMTG of the left hemisphere. This work shed light upon a language-general view of visual word reading across languages. As such, future studies are needed to disentangle task-related and script-specific demands in order to determine whether the involvement of regions, such as the left MFG, reflects general reading mechanisms or unique requirements of Chinese. Such nuanced investigations would allow researchers to clarify the extent to which Chinese reading engages specialized neural resources. In doing so, they could contribute to a more precise understanding of both universal and language-specific principles in the neuroscience of reading.

Despite the divergence between language-general and language-specific accounts regarding word reading, widely distributed brain regions have been identified to be associated with the processing of the basic linguistic components (e.g., word forms, word meaning). The neural mechanisms of the basic processing, as Zhao et al. (2017) pointed out, are extremely important for a better understanding of how a universal network adapted to reading words across languages is shaped by the unique properties of specific writing systems. However, most of the previously mentioned neuroimaging studies (e.g., Bolger et al. 2005) either failed to dissociate phonological, orthographic, and semantic processing, or only investigated one component of word processing (Hu et al. 2018; Tan et al. 2005a; Zhang et al. 2019; Zhao et al. 2017). More empirical studies or systematic reviews are in need to provide a more comprehensive picture of the brain patterns and changes in Chinese word reading. To address this gap, the following section reviews neuroimaging evidence on the temporal and spatial unfolding of both word form and meaning processing respectively.

2.3 Phonological, Orthographic, and Semantic Activation

2.3.1 Temporal Signatures

Visual word reading implicates a multi-stage process and simultaneously an anatomically distributed neural system (Carreiras et al. 2014; Liu et al. 2003). This system progressively aggregates the lower level of visual analysis to higher-level semantic regions, and links this information to word form (orthographic and/or phonological information) and meaning systems (Lo et al. 2019; Wu et al. 2012). The high temporal resolution of electroencephalography (EEG) technology provides a viable tool to investigate the time courses of phonological, orthographic, and semantic activation during word reading. Event-related potential (ERP) components extracted from EEG, such as N170, P200, and N400,[1] have been identified as the

[1] N170, P200, and N400 are ERP components that reflect specific cognitive processes. "N" and "P" indicate the polarity of the component (negative or positive), and the number refers to the approxi-

potential bio-markers of orthographic/visual analysis, phonological/orthographic, and semantic processing respectively in both alphabetic (Barnea and Breznitz 1998; Hauk et al. 2006) and non-alphabetic reading (Yum et al. 2014) (see related Chinese studies in Table 2.1). In the pioneer work from Liu et al. (2003), graphically related Chinese word pairs generated a smaller P200 in the pronunciation task and a smaller N400 in the meaning task, while phonologically related pairs elicited decreased N400 effect with bilateral resources in the meaning task. Their results depicted the temporal unfolding from graphic and phonological to semantic information in order, which was also addressed in the lexical constituency model (Perfetti et al. 2005). Meanwhile, their evidence illuminated the sensitivity of P200 and N400 to task demands of word reading process, which is informative to the following time-course studies.

Zhang et al. (2009) investigated the temporal dynamics of phonological and semantic activation during Chinese word reading. By employing semantic judgment and homophone judgment tasks with single-character stimuli, they identified a P200 effect for homophonic pairs, which was interpreted as evidence for an early phonological activation, preceding a frequency-sensitive N400 effect observed between semantically or phonologically related words and unrelated controls. These findings provided important insights into the temporal properties of Chinese lexical representation, particularly for single-character word processing. Notably, a more recent study by Wu et al. (2020) offered partial replication of these results. Using Chinese homographic morphemes, Wu and colleagues sought to disentangle orthographic, phonological, and semantic contributions. Their results suggested that early stages of visual word recognition are more strongly associated with morpho-orthographic and phonological processing at the lexeme level, whereas later stages engage morpho-semantic processing at the lemma level.

Moreover, in studies of alphabetic reading, the N170 effect has been consistently observed at occipito-temporal scalp sites, reflecting differences between orthographic and non-orthographic stimuli (Bentin et al. 1999). In the context of Chinese word reading, the N170 has similarly been interpreted as a marker of orthographic sensitivity, indexing early perceptual categorization or the initial stage of orthographic processing for Chinese characters (Hsu et al. 2009; Su et al. 2012; Yum et al. 2014, 2016; X. Zhang et al. 2020b).

In addition to N170 and P200, N200 is also recognized to be associated with Chinese orthographic processing. For instance, Zhang et al. (2012a) used a visual lexical decision task to differentiate the brain activation difference between Chinese words vs. pseudowords and Chinese vs. Korean words. In particular, they reported a centro-parietal N200, which is specific to Chinese visual word reading. This early effect was replicated by a handful of studies (Du et al. 2013; Gao et al. 2021; Zhou et al. 2016) and believed to implicate the early identification of the Chinese disyllabic word-form orthography involving extensive and advanced visual analysis.

---

mate latency of the peak in milliseconds. For a detailed introduction to EEG and ERP, see Sect. 4.3.1.

Table 2.1 Representative neuroimaging studies on Chinese character and word reading

Source	Task	Imaging tools	Linguistic stimuli	Main findings
Tan et al. (2000)	Word association	fMRI	Chinese single-character and disyllabic words	Strong left lateralization of frontal region (BAs 9 and 47)
Liu and Perfetti (2003)	Delayed naming	ERP source localization	Chinese characters and English words	Bilateral occipital region (left BA 17, right BA 18) for visual processing of Chinese characters
Liu et al. (2003)	Semantic and phonological judgment	ERP	Graphically similar, homophonic, semantically related word pairs	P200 phonological effect and N400 semantic effect
Kuo et al. (2004)	Homophone judgment and physical comparisons	fMRI	Chinese characters, pseudocharacters	The left occipitotemporal region for orthographic processing and left premotor gyrus, left middle/inferior frontal gyrus for phonological processing
Chen et al. (2008)	Homophone judgment	fNIRS	Chinese characters, English words	Activation in left middle frontal, left superior temporal, and left supramarginal gyri
Zhang et al. (2009)	Semantic and homophone judgment	ERP	Semantically and phonologically related word pairs	N400 semantic effect and P200 phonological effect
Chung et al. (2010)	Lexical decision	ERP	Chinese compound words	P250 effect for morphological structure processing
Wang (2011)	Stroop	ERP	Chinese single-character words	N450 effect for incongruent conditions
Zhang et al. (2012a)	Lexical decision	ERP	Real Chinese words and pseudo- or non-words	N200 orthographic effect

(continued)

Table 2.1 (continued)

Source	Task	Imaging tools	Linguistic stimuli	Main findings
Zhang et al. (2012b)	Lexical decision	fMRI	Chinese disyllabic words and nonwords	BA 44/45 in phonological processing, and BA 47 in semantic processing
Du et al. (2013)	Delayed repetition priming	ERP	Chinese disyllabic words and nonwords	N200 and N400 effects for morphological processing
Bulut et al. (2017)	Passive reading	fMRI	Chinese sentences, unstructured word lists, and disconnected character lists	Stronger activation in left IFG and left inferior parietal lobule in reading unstructured word lists
Hu et al. (2018)	Lexical decision	fNIRS	Chinese characters and pseudo-characters	Brain activation and connectivity in occipito-temporal cortex
Zhang et al. (2019)	Semantic judgment	fMRI	Chinese disyllabic words	Left IFG and left MTG for semantic processing
Ip et al. (2019)	Morphological and phonological judgment	fMRI	Two- and three-character words and pseudowords	Left temporal regions for morphological processing
Hsu et al. (2019)	Lexical decision	Magnetoencephalography (MEG)	Chinese disyllabic words	Left temporal cortex activity
Wu et al. (2020)	Masked priming and lexical decision	ERP	Homographic morphemes	P200 word form effect and N400 semantic effect
H. Zhang et al. (2020a)	Picture-character matching	ERP	Chinese characters	N170, P200, and N400 effects for orthographic and semantic activation
Wei et al. (2024)	Orthographic, phonological, semantic judgments	fMRI	Chinese characters	Right ventral occipitotemporal cortex (vOTC) for visuospatial orthographic processing

To summarize, evidence on the temporal dynamics of form-to-meaning processing has shown a relatively consistent pattern across different stimulus types and experimental paradigms. Overall, findings suggest that Chinese word reading follows a general temporal sequence, beginning with early orthographic and/or

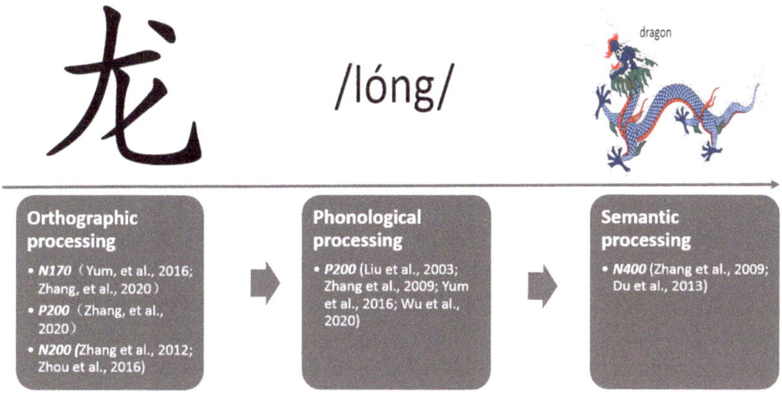

Fig. 2.1 Schematic illustration of the time course of orthographic, phonological, and semantic processing in Chinese word reading based on ERP studies

phonological processes (N170, P200, e.g., Wu et al. 2020; H. Zhang et al. 2020a) and followed by later semantic processing (N400) (see Fig. 2.1), a pattern broadly comparable to that reported for alphabetic scripts. At the same time, an additional orthographic marker has been reported in Chinese reading, namely the centro-parietal N200, which appears to reflect script-specific demands of the Chinese writing system.

2.3.2 Anatomical Distributions

Having reviewed the temporal dynamics of orthographic, phonological, and semantic processing in Chinese word reading, we now turn to questions of spatial localization. To map these processes onto the human cortex, psycholinguists and neurolinguists have manipulated a range of lexical variables and employed neuroimaging techniques with higher spatial resolution (e.g., fNIRS, fMRI, and MEG[2]) (see Table 2.1). Tan et al. (2005a), for instance, used an activation likelihood estimation (ALE) approach to synthesize brain activation data from six studies on Chinese word reading and compared the results with alphabetic languages. Their analysis revealed four neural systems implicated in phonological processing during Chinese reading: the left dorsolateral frontal region (BA 9), the dorsal aspect of the left inferior parietal lobule, a bilateral ventral occipito-temporal network (including the fusiform and middle occipital gyri), and a left ventral prefrontal system. Within this framework, Tan and colleagues proposed that the left MFG is specialized for the addressed phonology in Chinese, whereas the left temporo-parietal regions mediate the assembled phonology in alphabetic languages. Booth et al. (2006) further

[2] See Sections 4.3.2 to 4.3.4 for the introduction of fNIRS, fMRI, and MEG.

contrasted phonological and semantic processing through parallel rhyming and meaning-association tasks. Their results were broadly consistent with previous findings in English (Blumenfeld et al. 2006; Booth et al. 2002), showing that inferior frontal activations support the access and manipulation of phonological and semantic information, but also demonstrating additional recruitment of the MFG in Chinese.

Building on the work of Tan et al. (2005b) and Bolger et al. (2005), Wu et al. (2012) conducted a meta-analysis to delineate the neural networks underlying orthographic, phonological, and semantic processing in Chinese word reading. Their synthesis indicated that phonological processing engages the left inferior parietal lobule and the right STG, semantic processing relies primarily on the left MTG, and orthographic processing is supported by occipito-temporal regions. Moreover, a functional dissociation between phonological and semantic operations was observed within the left IFG. These results were largely consistent with earlier findings (Bolger et al. 2005; Cao et al. 2010; Tan et al. 2005a). At the same time, Wu and colleagues also reported additional recruitment of the left MFG and the right fusiform gyrus, highlighting the unique features of Chinese word reading that extend beyond the language-general networks established in alphabetic scripts.

More recently, Wei et al. (2024) refined this understanding through an fMRI study of visuospatial orthographic processing in Chinese single-character reading. They showed that, beyond the left ventral occipito-temporal cortex (vOTC), the right vOTC also plays a critical role, reflecting the spatially complex layout of Chinese characters. Importantly, functional connectivity analyses revealed stronger interactions among the left vOTC, superior parietal gyrus (SPG), and MFG during orthographic tasks than semantic ones, reinforcing the notion that Chinese reading requires additional visuospatial support relative to alphabetic systems.

Overall, current evidence indicates that Chinese orthographic processing engages occipito-temporal regions with bilateral activation patterns, a profile that stands in contrast to findings from alphabetic languages. For semantic processing, converging results show the involvement of left temporal and frontal cortices, largely consistent with observations in alphabetic reading. Variability in the specific loci of Chinese semantic effects appears to reflect the particular characteristics of the Chinese script and may also depend on task requirements. With regard to phonology, studies have demonstrated that the left dorsolateral frontal cortex and left parietal regions are recruited when tasks explicitly require phonological access. Yet, it remains unresolved whether Chinese word reading relies on uniquely specialized neural regions, underscoring the need for further systematic research.

2.3.3 Single vs. Dual Routes to Word Meaning

A fundamental issue in reading science is to understand the processes of meaning access from print. Dual routes model (Coltheart et al. 2001; Fiebach et al. 2002) argued that there exist two parallel routes in word reading (see Fig. 2.2). The lexical

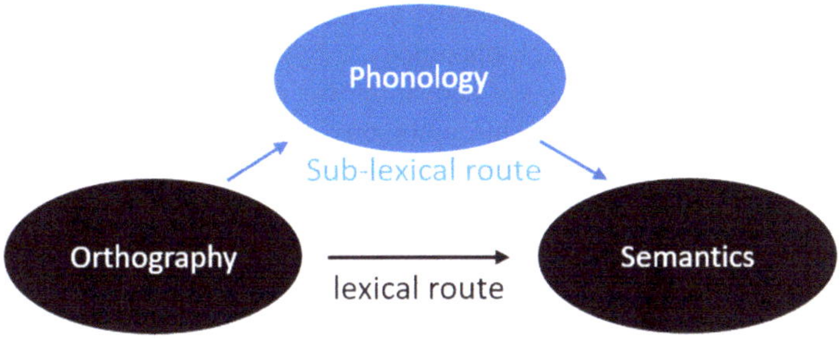

Fig. 2.2 Schematic demonstration of dual routes to word meaning

route enables direct mapping from orthographic forms to the semantic representations stored in the mental lexicon, whereas the sub-lexical route relies on grapheme-to-phoneme conversion to reach meaning. Although a large body of studies has demonstrated that phonological activation occurs during Chinese character recognition (e.g., Perfetti 2003; Perfetti et al. 2005), some researchers contend that phonology may not play a mandatory role in lexical access, given the high degree of opacity in form–sound correspondences within the Chinese writing system (Chen and Shu 2001; McBride-Chang et al. 2005; Tsang et al. 2018). The debate has been revitalized by neuroimaging evidence suggesting that phonology plays, at most, a limited role in Chinese word reading (Wang et al. 2010; Wong et al. 2014; Zhang et al. 2012b). In particular, Zhang et al. (2012b) employed a lexical decision paradigm with fMRI, manipulating both lexicality (words vs. nonwords) and modality (auditory vs. visual). Their findings showed that in the visual condition, nonwords elicited stronger activation than words in both BA 44/45 and BA 47, whereas in the auditory condition, greater nonword-related activation was confined to BA 44/45. Given that BA 44/45 has been linked to phonological processing and BA 47 to semantic processing (Bookheimer 2002), these results were taken to indicate that Chinese words are capable of directly activating semantic representations without necessarily relying on an indirect phonological route.

Converging ERP evidence (Wong et al. 2014) further suggested that phonological activation may be minimal in Chinese word reading. In their lexical decision task, no phonological effect (e.g., Ashby 2010; Ashby and Martin 2008) was observed in the syllable-related priming condition (e.g., 机场/ji1 chang3/, "airport"—基础/ji1 chu3/, "basics"). Instead, an N250 effect emerged in the character-related condition (e.g., 机场—机会/ji1 hui4/, "chance"), and an N400 effect was elicited in the semantic-related condition (e.g., 机场, "airport"—跑道/pao3 dao4/, "runway"). Based on the absence of phonological effects, the authors argued that phonological activation is not obligatory and that its role in processing Chinese two-character compounds could be minimized.

Nevertheless, a critical limitation lies in their design. The syllable-related condition was not a pure index of phonological association, since all stimuli were presented visually. Thus, the manipulation was inevitably confounded by orthographic factors (e.g., visual complexity) and morphological information (e.g., distinct morphological structures). Consequently, the null phonological effect may reflect either (a) a lexical route in which phonology is bypassed altogether, or (b) a sub-lexical route in which phonological activation was overshadowed by orthographic and morphological influences. This ambiguity underscores the need for finer-grained paradigms that can disentangle phonological contributions from other sources of variance. Importantly, this pattern stands in contrast to findings from alphabetic languages, where phonological mediation is typically considered central to lexical access.

Moreover, as noted by Zhang et al. (2012b), both their study and that of Wong et al. relied on a form of reasoning known as "reverse inference", whereby brain activation patterns previously linked to a given cognitive function are taken to imply the engagement of that function in new contexts. Although such inferences are not deductively valid, they can still yield limited yet useful insights (Poldrack 2006). Hence, it would be premature to dismiss the potential contribution of phonology to meaning access in Chinese word reading. Further research, with more rigorous experimental designs and analytic approaches, is clearly required before firm conclusions can be drawn regarding single-route versus dual-route models.

2.4 Lexical Variables, Individual Differences, and Social-Emotional Dimensions

Previous psycholinguistic research on word reading and lexical processing has primarily concentrated on two variables: frequency and semantic transparency. Frequency includes both word-level and morpheme/character-level frequency, whereas semantic transparency describes the degree to which the meaning of individual morphemes corresponds to the overall meaning of the word. Behavioral studies have consistently shown that high-frequency words or words with lower transparency elicit faster responses than their low-frequency or more transparent counterparts when other lexical properties are controlled (Sandra 1994; Wang et al. 1999). With the advancement of neuroimaging techniques such as EEG and fMRI, these behavioral findings have been extended to neural evidence. For example, Kuo et al. (2004), using an event-related fMRI paradigm, demonstrated that low-frequency characters elicited stronger activation in the left IFG, supplementary motor area, left anterior insula, left posterior temporal gyrus (PTG), left superior parietal cortex, and lingual gyrus, whereas high-frequency characters recruited the left supramarginal/angular gyrus and precuneus. Consistent evidence was reported by Lin et al. (2016), who found that frequency effects were reflected in the bilateral inferior parietal lobules and posterior cingulate cortex. Taken together, these findings suggest that frequency modulates processes of semantic integration and retrieval in Chinese reading.

Temporal aspects of frequency effects have also been identified in ERP research on Chinese word reading. Zhang and Peng (1992) firstly reported that, when word frequency was controlled, the frequencies of the two constitute morphemes of coordinate words determined the reaction times to the stimuli, while only the second morpheme's frequency affected the latency to subordinate words. This indicates that frequency is a useful index for examining how sub-lexical information shapes lexical access during reading. To further trace its time course in the brain, Zhang et al. (2009) compared ERPs elicited by high- and low-frequency single-character word pairs in semantic and homophone judgment tasks. Their results revealed a word frequency effect on the N400, but no differences at earlier ERP components. Nonetheless, most ERP studies so far have concentrated on word-level frequency (Gu and Yang 2010; Liu and Perfetti 2003; Qu et al. 2016), leaving the influence of morpheme frequency on lexical-semantic access underexplored. More systematic investigations are therefore warranted in this domain.

Meanwhile, although the effect of semantic transparency has been extensively investigated with neuroimaging methods in alphabetic languages (e.g., Fiorentino et al. 2015; Lavric et al. 2012; Simon et al. 2012; Solomyak and Marantz 2009), relatively fewer studies have systematically examined its temporal and spatial correlates in Chinese word reading. In recent years, however, an increasing number of studies have begun to address the psychological and neural reality of semantic transparency in Chinese, though the findings remain somewhat mixed. For instance, recent work has demonstrated its varying degrees of influence on visual and auditory compound word processing drawing on ERP and fMRI techniques (Han et al. 2025; Lee et al. 2021; Tsang et al. 2022; Wei et al. 2023; Zhang et al. 2024). Moving forward, neuroimaging research that integrates these multimodal perspectives and systematically examines how semantic transparency interacts with other lexical variables will be essential. Clarifying its neural basis in Chinese will not only advance our understanding of the specific mechanisms underlying Chinese reading but also enrich cross-linguistic models of morphological processing.

Interestingly, an increasing number of studies over the past two decades have focused on the role of word category information. Word category can be defined in terms of grammatical class (e.g., nouns vs. verbs) or semantic class (e.g., objects vs. actions, animate vs. inanimate). According to the speech production model (Levelt 1989), the mental lexicon contains an independent syntactic representation that interacts with semantic information at the higher lemma level of lexical encoding. This suggests that a lexical entry's syntactic properties, such as grammatical category, are also activated during lexical access and word production. Accumulating evidence suggests distinct neurocognitive distributions for word category (e.g., the left STG for nouns and the left IFG for verbs) from both alphabetic languages (e.g., Davis et al. 2004; Momenian et al. 2016; Palti et al. 2007; Vigliocco et al. 2011) and Chinese (Li et al. 2004; Yang et al. 2017, 2011, 2002; Yu et al. 2011). However, the noun vs. verb difference in the human brain remains controversial. In particular, it is unclear whether nouns and verbs are dissociable from each other as grammatical or semantic categories. The debate is even more pronounced for Chinese word reading, in which there is no morphological marker for grammatical inflections. Xia et al. (2016) manipulated the word class (noun vs. verb) and number of syllables (monosyllabic vs. disyllabic) in a semantic relatedness judgment task with EEG recordings. All four types of words were embedded in syntactically well-defined contexts, such as 一个岛 (one **island**), 一个书包 (one **bag**), 不该搬 (shouldn't **move**), and 不该迎接 (shouldn't **greet**) for monosyllabic noun, disyllabic noun, monosyllabic verb, and disyllabic verb, respectively. The findings revealed that Chinese verbs elicited a larger N400 response than nouns in both monosyllabic and disyllabic contexts. Given that the N400 is associated with lexical access and semantic memory retrieval, this noun–verb contrast in Chinese is more likely attributable to differences in conceptual meaning between objects and actions rather than to grammatical category distinctions.

In contrast, a more recent fMRI study reported the opposite pattern and interpretation. Feng et al. (2020) compared nouns and verbs in both syntactic sessions (e.g., 一位老师, a **teacher** vs. 一味宽容, blindly **tolerate**) and semantic sessions (e.g., 杰出人才, distinguished **talents** vs. 充分强调, fully **emphasize**). Their results showed that verbs elicited stronger activation than nouns in the left frontal lobe during semantic processing, with more widespread recruitment across frontal and temporal regions. In the syntactic task, verbs also produced stronger activation than nouns, particularly in the left IFG, consistent with prior findings in Indo-European languages. These results suggest that the brain may differentiate nouns and verbs in Chinese by relying more heavily on grammatical rules than on semantic distinctions, reflecting an economical neural strategy. The discrepancy between this and earlier ERP evidence may stem from differences in task design and methodological emphasis. For example, Xia et al. (2016) employed a semantic relatedness paradigm, where syntactic processing was not directly engaged. To reconcile these inconsistencies, further research is needed to test whether these patterns generalize across diverse semantic and syntactic tasks and word contexts. Moreover, combining complementary techniques (e.g., EEG–fMRI or EEG–fNIRS) will also be valuable for advancing our understanding of how Chinese nouns and verbs are represented in the brain. Such findings will not only clarify the neural basis of noun–verb distinctions in Chinese but also contribute to cross-linguistic theories of lexical category representation.

Notably, a major development in reading research in the new millennium has been the increasing adoption of an individual-differences perspective, which highlights variability across readers rather than relying exclusively on group-level generalizations. Variables of interest span a wide range, including age (Siok et al. 2020), dyslexia (Han et al. 2005; Lee et al. 2005; Liu et al. 2013) and children (Ji et al. 2019; Lo et al. 2019; Takashima et al. 2019; Wulff et al. 2019) population, cognitive constraints such as working memory (Buchweitz et al. 2009), and second language or bilingual contexts (Liu and Cao 2016; Sulpizio et al. 2020; Tagarelli et al. 2019; Yu et al. 2019; X. Zhang et al. 2020b), among others. In particular, Siok et al. (2020) conducted a large-scale cross-sectional fMRI study with 125 Chinese native speakers aged 6 to 74 years, providing the first lifespan account of Chinese reading development. Their results indicated that a general network encompassing the left frontal and occipital cortices supports Chinese reading across age groups, but activations in the left mid-inferior frontal gyrus and fusiform gyrus gradually decreased with age. Crucially, the cortical network for Chinese reading differs from that of alphabetic systems, and these differences are evident even at the earliest stages of literacy acquisition.

This line of research not only deepens our understanding of how learning experiences and biological maturation jointly shape the reading brain, but also offers a valid approach for investigating lifespan trajectories of brain development. Moreover, the individual-differences approach underscores that reading is not a uniform process, and insights gained from Chinese reading research can refine cross-linguistic models of literacy by accounting for variability across populations, contexts, and developmental stages. In addition, such findings provide important implications for educational practice, including literacy instruction across different age groups and the design of targeted interventions for individuals with reading difficulties.

Beyond individual differences, socio-emotional factors also play an essential role in shaping the reading brain. Although the neurocognitive study of reading primarily focuses on the brain mechanisms engaged during word recognition and comprehension, it would be misleading to overlook the broader sociocultural and emotional dimensions of reading behavior. From the perspective of distributed cognition, mental processes are not confined to the individual mind but are inherently embedded in interactions between agents and their environments (Ellis 2019; Hutchins 1995).

A growing body of research has begun to explore the influence of emotion on word reading. ERP studies, for instance, have shown that emotional meaning can modulate both early perceptual stages and later semantic integration during word reading (Herbert et al. 2008; Xu et al. 2019), and further confirmed the psychological reality of distinguishing between *emotion-label words* (e.g., 开心/kai1 xin1/, "happy") and *emotion-laden words* (e.g., 婚礼/hun1 li3/, "wedding") in Chinese context (Zhang et al. 2017, 2018). Cross-cultural neuroimaging evidence further revealed that emotion is intertwined with cultural background. Chen et al. (2019) found that, compared with Western participants, Eastern readers exhibited stronger activation in the left cerebellum and thalamus when processing positive words, and in the right precuneus and left superior parietal lobule when processing negative words, suggesting that cultural context shapes emotional regulation during reading.

Socio-cultural environments also leave their mark on reading development. In Chinese, for example, orthographic skills are facilitated by handwriting practice (Chan et al. 2006; Tan et al. 2005b) but can be undermined by reliance on electronic input systems (Tan et al. 2013; Zhou et al. 2020). Zhou et al. (2020) reported that children who frequently used pinyin input exhibited weaker cortical activation networks in regions such as the left MFG, left IFG, and right fusiform gyrus, compared with peers who relied less on pinyin. Such findings highlight both the opportunities and challenges for children's neurodevelopment in the digital age and provide valuable insights for language policy decisions.

Nevertheless, systematic investigations into the socio-emotional dimensions of reading remain relatively scarce, with most progress occurring only in the last decade. Advancing reading science therefore requires a more sustained effort to integrate sociocultural and affective perspectives alongside cognitive and neural approaches.

2.5 Summary

Research on Chinese reading in the new millennium has highlighted both universal and language-specific characteristics of word recognition. From a typological perspective, Chinese contrasts maximally with alphabetic languages: it employs a logographic script where characters correspond to morpheme-syllable units rather than phonemes, and most words are disyllabic compounds rather than inflected forms. These linguistic properties underpin reading processes that diverge from those of alphabetic systems. Neuroimaging studies have revealed that while Chinese reading engages core language-general networks, such as the left IFG for semantic access, it also recruits distinct regions including the left MFG and the right fusiform gyrus, reflecting the visuospatial and addressed-phonology demands of Chinese script.

Moreover, electrophysiological and neuroimaging evidence on phonological, orthographic, and semantic activation suggests both convergences and divergences between Chinese and alphabetic reading. Orthographic processing tends to involve bilateral occipito-temporal regions, semantic processing relies heavily on left-lateralized temporal and frontal areas, and phonological processing shows task-dependent routes, fueling ongoing debates about single- versus dual-route accounts of meaning access. Recent work has also expanded to include lexical variables (e.g., frequency, transparency) and individual differences (e.g., age, socio-cultural environment), underscoring the dynamic interplay between cognitive mechanisms and reader variability. Collectively, these findings indicate that a full account of Chinese reading requires integrating both universal cognitive principles and script-specific constraints.

In this chapter, we have further argued that several issues remain unresolved. First, the precise anatomical specialization for orthographic, phonological, and semantic processing in Chinese remains unclear, partly due to task-specific confounds and modality effects. Second, the debate on whether phonological mediation is necessary in Chinese word recognition highlights the need for replication studies with improved designs to test single-route versus dual-route accounts. Third, methodological innovation is required: statistical advances such as Bayesian inference and meta-analytic synthesis, as well as multimodal neuroimaging (e.g., EEG-fMRI), can offer more robust insights into the neural architecture of reading. Importantly, the neural underpinnings of morphological processing, especially how readers extract and utilize morphological structure, remain underexplored despite the centrality of morphology in Chinese. Addressing this gap will not only illuminate language-specific mechanisms but also contribute to broader theories of reading.

Building on these insights, Chap. 3 turns to morphology and morphological processing, one of the most distinctive features of the Chinese language. Since Chinese words are largely compounds without inflectional morphology, understanding how readers process morphemes and combine them into words is essential for revealing both script-specific adaptations and universal principles of lexical organization. The following section reviews behavioral and neuroimaging studies on Chinese morphology, with a particular focus on how morphological structure shapes word recognition, semantic integration, and reading fluency.

References

Anderson, R. C., & Chen, X. (2013). Chinese Reading Development in Monolingual and Bilingual Learners: Introduction to the Special Issue. *Scientific Studies of Reading*, *17*(1), 1-4. https://doi.org/10.1080/10888438.2012.729120

Ashby, J. (2010). Phonology is fundamental in skilled reading: Evidence from ERPs. *Psychonomic Bulletin & Review*, *17*(1), 95-100.

Ashby, J., & Martin, A. E. (2008). Prosodic phonological representations early in visual word recognition. *Journal of Experimental Psychology: Human perception and performance*, *34*(1), 224.

Barnea, A., & Breznitz, Z. (1998). Phonological and orthographic processing of Hebrew words: electrophysiological aspects. *The Journal of Genetic Psychology*, *159*(4), 492-504.

Bentin, S., Mouchetant-Rostaing, Y., Giard, M.-H., Echallier, J.-F., & Pernier, J. (1999). ERP manifestations of processing printed words at different psycholinguistic levels: time course and scalp distribution. *Journal of Cognitive Neuroscience*, *11*(3), 235-260.

Blumenfeld, H. K., Booth, J. R., & Burman, D. D. (2006). Differential prefrontal–temporal neural correlates of semantic processing in children. *Brain and Language*, *99*(3), 226-235.

Bolger, D. J., Perfetti, C. A., & Schneider, W. (2005). Cross-cultural effect on the brain revisited: universal structures plus writing system variation. *Hum Brain Mapp*, *25*(1), 92-104. https://doi.org/10.1002/hbm.20124

Bookheimer, S. (2002). Functional MRI of language: new approaches to understanding the cortical organization of semantic processing. *Annual review of neuroscience*, *25*(1), 151-188.

Booth, J. R., Burman, D. D., Meyer, J. R., Gitelman, D. R., Parrish, T. B., & Mesulam, M. M. (2002). Modality independence of word comprehension. *Human brain mapping*, *16*(4), 251-261.

Booth, J. R., Lu, D., Burman, D. D., Chou, T.-L., Jin, Z., Peng, D.-L., Zhang, L., Ding, G.-S., Deng, Y., & Liu, L. (2006). Specialization of phonological and semantic processing in Chinese word reading. *Brain research*, *1071*(1), 197-207.

Buchweitz, A., Mason, R. A., Tomitch, L. M., & Just, M. A. (2009). Brain activation for reading and listening comprehension: An fMRI study of modality effects and individual differences in language comprehension. *Psychol Neurosci*, *2*(2), 111-123. https://doi.org/10.3922/j.psns.2009.2.003

Bulut, T., Hung, Y. H., Tzeng, O., & Wu, D. H. (2017). Neural correlates of processing sentences and compound words in Chinese. *PLoS One*, *12*(12), e0188526. https://doi.org/10.1371/journal.pone.0188526

Cain, K., & Parrila, R. (2013). Introduction to the Special Issue. Theories of Reading: What We Have Learned From Two Decades of Scientific Research. *Scientific Studies of Reading*, *18*(1), 1-4. https://doi.org/10.1080/10888438.2013.836525

Cao, F. (2015). Neuroimaging studies of reading in bilinguals. *Bilingualism: Language and Cognition*, *19*(4), 683-688. https://doi.org/10.1017/s1366728915000656

Cao, F., Lee, R., Shu, H., Yang, Y., Xu, G., Li, K., & Booth, J. R. (2010). Cultural constraints on brain development: evidence from a developmental study of visual word processing in mandarin chinese. *Cereb Cortex*, *20*(5), 1223-1233. https://doi.org/10.1093/cercor/bhp186

Carreiras, M., Armstrong, B. C., Perea, M., & Frost, R. (2014). The what, when, where, and how of visual word recognition. *Trends Cogn Sci*, *18*(2), 90-98. https://doi.org/10.1016/j.tics.2013.11.005

Chan, D. W., Ho, C. S.-H., Tsang, S.-M., Lee, S.-H., & Chung, K. K. (2006). Exploring the reading–writing connection in Chinese children with dyslexia in Hong Kong. *Reading and Writing*, *19*(6), 543-561.

Chen, C., Xue, G., Mei, L., Chen, C., & Dong, Q. (2009). Cultural neurolinguistics. *Progress in brain research*, *178*, 159-171.

Chen, H.-C., & Shu, H. (2001). Lexical activation during the recognition of Chinese characters: Evidence against early phonological activation. *Psychonomic Bulletin & Review*, *8*(3), 511-518.

Chen, H. C., Vaid, J., Bortfeld, H., & Boas, D. A. (2008). Optical imaging of phonological processing in two distinct orthographies. *Exp Brain Res*, *184*(3), 427-433. https://doi.org/10.1007/s00221-007-1200-0

Chen, P., Chen, B., Münte, T. F., Lu, C., Liu, L., & Guo, T. (2019). Neural correlates of processing emotions in words across cultures. *Journal of Neurolinguistics*, *51*, 111-120. https://doi.org/10.1016/j.jneuroling.2019.01.004

Chou, T.-L., Chen, C.-W., Wu, M.-Y., & Booth, J. R. (2009). The role of inferior frontal gyrus and inferior parietal lobule in semantic processing of Chinese characters. *Experimental brain research*, *198*(4), 465-475.

Chung, K. K., Tong, X., Liu, P. D., McBride-Chang, C., & Meng, X. (2010). The processing of morphological structure information in Chinese coordinative compounds: an event-related potential study. *Brain Res*, *1352*, 157-166. https://doi.org/10.1016/j.brainres.2010.06.069

Coltheart, M., Curtis, B., Atkins, P., & Haller, M. (1993). Models of reading aloud: Dual-route and parallel-distributed-processing approaches. *Psychological review*, *100*(4), 589.

Coltheart, M., Rastle, K., Perry, C., Langdon, R., & Ziegler, J. (2001). DRC: a dual route cascaded model of visual word recognition and reading aloud. *Psychological review*, *108*(1), 204.

Davis, M. H., Meunier, F., & Marslen-Wilson, W. D. (2004). Neural responses to morphological, syntactic, and semantic properties of single words: an fMRI study. *Brain and Language*, *89*(3), 439-449.

DeFrancis, J. (1989). *Visible speech: The diverse oneness of writing systems*. University of Hawaii Press.

Du, Y., Hu, W., Fang, Z., & Zhang, J. X. (2013). Electrophysiological correlates of morphological processing in Chinese compound word recognition. *Front Hum Neurosci*, *7*, 601. https://doi.org/10.3389/fnhum.2013.00601

Ellis, N. C. (2019). Essentials of a theory of language cognition. *The Modern Language Journal*, *103*, 39-60.

Feng, S., Qi, R., Yang, J., Yu, A., & Yang, Y. (2020). Neural correlates for nouns and verbs in phrases during syntactic and semantic processing: An fMRI study. *Journal of Neurolinguistics*, *53*. https://doi.org/10.1016/j.jneuroling.2019.100860

Fiebach, C. J., Friederici, A. D., Müller, K., & Von Cramon, D. Y. (2002). fMRI evidence for dual routes to the mental lexicon in visual word recognition. *Journal of Cognitive Neuroscience*, *14*(1), 11-23.

Fiorentino, R., Politzer-Ahles, S., Pak, N. S., Martínez-García, M. T., & Coughlin, C. (2015). Dissociating morphological and form priming with novel complex word primes: Evidence from masked priming, overt priming, and event-related potentials. *The Mental Lexicon*, *10*(3), 413-434.

Gao, F., Wang, J., Wu, C., Wang, M.-Y., Zhang, J., & Yuan, Z. (2021). The neural dynamics associated with lexicality effect in reading single Chinese words, pseudo-words and non-words. *Cognitive Neurodynamics*, 1-11.

Gibson, E. J., & Levin, H. (1975). *The psychology of reading*. The MIT press.

Gu, J., & Yang, Y. (2010). A neuro-electrophysiological study on productivity of Chinese compounding [in Chinese]. *Yuyan Wenzi Yingyong* (3), 98-107.

Han, Y.-J., Huang, S.-C., & Cheng, S.-K. (2025). ERP Evidence for the Modulation of Semantic Transparency on the Recollection of Two-Character Chinese Words. In *Cognitive and Neural Foundations of Chinese Reading: From Learning to Advanced Processing and Beyond* (pp. 175-197). Springer.

Han, Z., Bi, Y., Shu, H., & Weekes, B. S. (2005). The interaction between semantic and sublexical routes in reading: Converging evidence from Chinese. *Brain and Language*, *95*(1), 235-236. https://doi.org/10.1016/j.bandl.2005.07.120

Hauk, O., Davis, M. H., Ford, M., Pulvermuller, F., & Marslen-Wilson, W. D. (2006). The time course of visual word recognition as revealed by linear regression analysis of ERP data. *Neuroimage*, *30*(4), 1383-1400. https://doi.org/10.1016/j.neuroimage.2005.11.048

Herbert, C., Junghofer, M., & Kissler, J. (2008). Event related potentials to emotional adjectives during reading. *Psychophysiology*, *45*(3), 487-498. https://doi.org/10.1111/j.1469-8986.2007.00638.x

Hsu, C.-H., Pylkkänen, L., & Lee, C.-Y. (2019). Effects of morphological complexity in left temporal cortex: An MEG study of reading Chinese disyllabic words. *Journal of Neurolinguistics*, *49*, 168-177. https://doi.org/10.1016/j.jneuroling.2018.06.004

Hsu, C. H., Tsai, J. L., Lee, C. Y., & Tzeng, O. J. (2009). Orthographic combinability and phonological consistency effects in reading Chinese phonograms: an event-related potential study. *Brain Lang*, *108*(1), 56-66. https://doi.org/10.1016/j.bandl.2008.09.002

Hu, Z., Zhang, J., Couto, T. A., Xu, S., Luan, P., & Yuan, Z. (2018). Optical Mapping of Brain Activation and Connectivity in Occipitotemporal Cortex During Chinese Character Recognition. *Brain Topogr*, *31*(6), 1014-1028. https://doi.org/10.1007/s10548-018-0650-y

Huey, E. B. (1908). *The psychology and pedagogy of reading: With a review of the history of reading and writing and of methods, texts, and hygiene in reading*. Literary Licensing, LLC.

Hutchins, E. (1995). How a cockpit remembers its speeds. *Cognitive science*, *19*(3), 265-288.

Ip, K. I., Marks, R. A., Hsu, L. S., Desai, N., Kuan, J. L., Tardif, T., & Kovelman, L. (2019). Morphological processing in Chinese engages left temporal regions. *Brain Lang, 199*, 104696. https://doi.org/10.1016/j.bandl.2019.104696

Ji, Y., Liu, J., Zhu, X.-Q., Zhao, J., Wang, J., Du, Y.-C., & Bi, H.-Y. (2019). Atypical N170 lateralization of face and word recognition in Chinese children with autism spectrum disorder. *Journal of Neurolinguistics*, *52*. https://doi.org/10.1016/j.jneuroling.2019.100858

Jobard, G., Crivello, F., & Tzourio-Mazoyer, N. (2003). Evaluation of the dual route theory of reading: a metanalysis of 35 neuroimaging studies. *Neuroimage*, *20*(2), 693-712.

Kuo, W. J., Yeh, T. C., Lee, J. R., Chen, L. F., Lee, P. L., Chen, S. S., Ho, L. T., Hung, D. L., Tzeng, O. J., & Hsieh, J. C. (2004). Orthographic and phonological processing of Chinese characters: an fMRI study. *Neuroimage*, *21*(4), 1721-1731. https://doi.org/10.1016/j.neuroimage.2003.12.007

LaBerge, D., & Samuels, S. J. (1974). Toward a theory of automatic information processing in reading. *Cognitive psychology*, *6*(2), 293-323.

Lavric, A., Elchlepp, H., & Rastle, K. (2012). Tracking hierarchical processing in morphological decomposition with brain potentials. *Journal of Experimental Psychology: Human perception and performance*, *38*(4), 811.

Lee, C. L., Hung, D. L., Tse, J. K., Lee, C. Y., Tsai, J. L., & Tzeng, O. J. (2005). Processing of disyllabic compound words in Chinese aphasia: evidence for the processing limitations account. *Brain Lang*, *92*(2), 168-184. https://doi.org/10.1016/j.bandl.2004.06.002

Lee, H.-J., Cheng, S.-K., Lee, C.-Y., & Kuo, W.-J. (2021). The neural basis of compound word processing revealed by varying semantic transparency and morphemic neighborhood size. *Brain and Language*, *221*, 104985.

Levelt, W. J. (1989). *Speaking: From intention to articulation*. MIT Press.

Li, G., Cheung, R. T., Gao, J. H., Lee, T. M., Tan, L. H., Fox, P. T., Jack Jr, C. R., & Yang, E. S. (2006). Cognitive processing in Chinese literate and illiterate subjects: An fMRI study. *Human brain mapping*, *27*(2), 144-152.

Li, P., Jin, Z., & Tan, L. H. (2004). Neural representations of nouns and verbs in Chinese: an fMRI study. *Neuroimage*, *21*(4), 1533-1541.

Lin, N., Yu, X., Zhao, Y., & Zhang, M. (2016). Functional Anatomy of Recognition of Chinese Multi-Character Words: Convergent Evidence from Effects of Transposable Nonwords, Lexicality, and Word Frequency. *PLoS One*, *11*(2), e0149583. https://doi.org/10.1371/journal.pone.0149583

Liu, H., & Cao, F. (2016). L1 and L2 processing in the bilingual brain: A meta-analysis of neuroimaging studies. *Brain Lang*, *159*, 60-73. https://doi.org/10.1016/j.bandl.2016.05.013

Liu, L., Tao, R., Wang, W., You, W., Peng, D., & Booth, J. R. (2013). Chinese dyslexics show neural differences in morphological processing. *Dev Cogn Neurosci*, *6*, 40-50. https://doi.org/10.1016/j.dcn.2013.06.004

Liu, Y., & Perfetti, C. A. (2003). The time course of brain activity in reading English and Chinese: an ERP study of Chinese bilinguals. *Hum Brain Mapp*, *18*(3), 167-175. https://doi.org/10.1002/hbm.10090

Liu, Y., Perfetti, C. A., & Hart, L. (2003). ERP evidence for the time course of graphic, phonological, and semantic information in Chinese meaning and pronunciation decisions. *J Exp Psychol Learn Mem Cogn*, *29*(6), 1231-1247. https://doi.org/10.1037/0278-7393.29.6.1231

Lo, J. C. M., McBride, C., Ho, C. S., & Maurer, U. (2019). Event-related potentials during Chinese single-character and two-character word reading in children. *Brain Cogn*, *136*, 103589. https://doi.org/10.1016/j.bandc.2019.103589

McBride-Chang, C., Chow, B. W. Y., Zhong, Y., Burgess, S., & Hayward, W. G. (2005). Chinese character acquisition and visual skills in two Chinese scripts. *Reading and Writing*, *18*(2), 99-128. https://doi.org/10.1007/s11145-004-7343-5

Momenian, M., Nilipour, R., Samar, R. G., Oghabian, M. A., & Cappa, S. (2016). Neural correlates of verb and noun processing: An fMRI study of Persian. *Journal of Neurolinguistics*, *37*, 12-21.

Nakamura, K., Kuo, W. J., Pegado, F., Cohen, L., Tzeng, O. J., & Dehaene, S. (2012). Universal brain systems for recognizing word shapes and handwriting gestures during reading. *Proc Natl Acad Sci U S A*, *109*(50), 20762-20767. https://doi.org/10.1073/pnas.1217749109

Packard, J. L. (2000). *The morphology of Chinese: A linguistic and cognitive approach*. Cambridge University Press.

Palti, D., Ben Shachar, M., Hendler, T., & Hadar, U. (2007). Neural correlates of semantic and morphological processing of Hebrew nouns and verbs. *Human brain mapping*, *28*(4), 303-314.

Paulesu, E., McCrory, E., Fazio, F., Menoncello, L., Brunswick, N., Cappa, S. F., Cotelli, M., Cossu, G., Corte, F., & Lorusso, M. (2000). A cultural effect on brain function. *Nature neuroscience*, *3*(1), 91-96.

Perfetti, C., Cao, F., & Booth, J. (2013). Specialization and Universals in the Development of Reading Skill: How Chinese Research Informs a Universal Science of Reading. *Sci Stud Read*, *17*(1), 5-21. https://doi.org/10.1080/10888438.2012.689786

Perfetti, C., & Stafura, J. (2013). Word Knowledge in a Theory of Reading Comprehension. *Scientific Studies of Reading*, *18*(1), 22-37. https://doi.org/10.1080/10888438.2013.827687

Perfetti, C. A. (1985). *Reading ability*. Oxford University Press.

Perfetti, C. A. (2003). The universal grammar of reading. *Scientific Studies of Reading*, *7*(1), 3-24.

Perfetti, C. A., Liu, Y., & Tan, L. H. (2005). The lexical constituency model: some implications of research on Chinese for general theories of reading. *Psychol Rev*, *112*(1), 43-59. https://doi.org/10.1037/0033-295X.112.1.43

Poldrack, R. A. (2006). Can cognitive processes be inferred from neuroimaging data? *Trends in cognitive sciences*, *10*(2), 59-63.

Price, C. J. (2000). The anatomy of language: contributions from functional neuroimaging. *The Journal of Anatomy*, *197*(3), 335-359.

Pugh, K. R., Mencl, W. E., Jenner, A. R., Katz, L., Frost, S. J., Lee, J. R., Shaywitz, S. E., & Shaywitz, B. A. (2000). Functional neuroimaging studies of reading and reading disability (developmental dyslexia). *Mental retardation and developmental disabilities research reviews*, *6*(3), 207-213.

Qu, Q., Zhang, Q., & Damian, M. F. (2016). Tracking the time course of lexical access in orthographic production: An event-related potential study of word frequency effects in written picture naming. *Brain and Language*, *159*, 118-126.

Rueckl, J. G., Paz-Alonso, P. M., Molfese, P. J., Kuo, W. J., Bick, A., Frost, S. J., Hancock, R., Wu, D. H., Mencl, W. E., Dunabeitia, J. A., Lee, J. R., Oliver, M., Zevin, J. D., Hoeft, F., Carreiras, M., Tzeng, O. J., Pugh, K. R., & Frost, R. (2015). Universal brain signature of proficient reading: Evidence from four contrasting languages. *Proc Natl Acad Sci U S A*, *112*(50), 15510-15515. https://doi.org/10.1073/pnas.1509321112

Sandra, D. (1994). The morphology of the mental lexicon: Internal word structure viewed from a psycholinguistic perspective. *Language and Cognitive Processes*, *9*(3), 227-269.

Simon, D. A., Lewis, G., & Marantz, A. (2012). Disambiguating form and lexical frequency effects in MEG responses using homonyms. *Language and Cognitive Processes*, *27*(2), 275-287.

Siok, W. T., Jia, F., Liu, C. Y., Perfetti, C. A., & Tan, L. H. (2020). A Lifespan fMRI Study of Neurodevelopment Associated with Reading Chinese. *Cereb Cortex*, *30*(7), 4140-4157. https://doi.org/10.1093/cercor/bhaa038

Solomyak, O., & Marantz, A. (2009). Lexical access in early stages of visual word processing: A single-trial correlational MEG study of heteronym recognition. *Brain and Language*, *108*(3), 191-196.

Su, I.-F., Mak, S.-C. C., Cheung, L.-Y. M., & Law, S.-P. (2012). Taking a radical position: evidence for position-specific radical representations in Chinese character recognition using masked priming ERP. *Frontiers in psychology*, *3*, 333.

Sulpizio, S., Del Maschio, N., Fedeli, D., & Abutalebi, J. (2020). Bilingual language processing: A meta-analysis of functional neuroimaging studies. *Neurosci Biobehav Rev*, *108*, 834-853. https://doi.org/10.1016/j.neubiorev.2019.12.014

Tagarelli, K. M., Shattuck, K. F., Turkeltaub, P. E., & Ullman, M. T. (2019). Language learning in the adult brain: A neuroanatomical meta-analysis of lexical and grammatical learning. *Neuroimage*, *193*, 178-200. https://doi.org/10.1016/j.neuroimage.2019.02.061

Takashima, A., Bakker-Marshall, I., van Hell, J. G., McQueen, J. M., & Janzen, G. (2019). Neural correlates of word learning in children. *Dev Cogn Neurosci*, *37*, 100649. https://doi.org/10.1016/j.dcn.2019.100649

Tan, L. H., Laird, A. R., Li, K., & Fox, P. T. (2005a). Neuroanatomical correlates of phonological processing of Chinese characters and alphabetic words: a meta-analysis. *Hum Brain Mapp*, *25*(1), 83-91. https://doi.org/10.1002/hbm.20134

Tan, L. H., & Li, H. (2016). Introduction: Brain mapping studies on Chinese cognitive processing [in Chinese]. *Contemporary Linguistics*, *18*(4), 544-548.

Tan, L. H., Liu, H.-L., Perfetti, C. A., Spinks, J. A., Fox, P. T., & Gao, J.-H. (2001). The neural system underlying Chinese logograph reading. *Neuroimage*, *13*(5), 836-846.

Tan, L. H., Spinks, J. A., Eden, G. F., Perfetti, C. A., & Siok, W. T. (2005b). Reading depends on writing, in Chinese. *Proceedings of the National Academy of Sciences*, *102*(24), 8781-8785.

Tan, L. H., Spinks, J. A., Gao, J. H., Liu, H. L., Perfetti, C. A., Xiong, J., Stofer, K. A., Pu, Y., Liu, Y., & Fox, P. T. (2000). Brain activation in the processing of Chinese characters and words: a functional MRI study. *Human brain mapping*, *10*(1), 16-27.

Tan, L. H., Xu, M., Chang, C. Q., & Siok, W. T. (2013). China's language input system in the digital age affects children's reading development. *Proc Natl Acad Sci U S A*, *110*(3), 1119-1123. https://doi.org/10.1073/pnas.1213586110

Taylor, I., & Taylor, M. M. (2014). *Writing and literacy in Chinese, Korean and Japanese*. John Benjamins Publishing Company.

Thorndike, E. L. (1917). Reading as reasoning: A study of mistakes in paragraph reading. *Journal of Educational Psychology*, *8*(6), 323.

Tsang, Y.-K., Zou, Y., & Tse, C.-Y. (2022). Semantic transparency in Chinese compound word processing: Evidence from mismatch negativity. *Neuroscience*, *490*, 216-223.

Tsang, Y. K., Huang, J., Lui, M., Xue, M., Chan, Y. F., Wang, S., & Chen, H. C. (2018). MELD-SCH: A megastudy of lexical decision in simplified Chinese. *Behav Res Methods*, *50*(5), 1763-1777. https://doi.org/10.3758/s13428-017-0944-0

Turkeltaub, P. E., Eden, G. F., Jones, K. M., & Zeffiro, T. A. (2002). Meta-analysis of the functional neuroanatomy of single-word reading: method and validation. *Neuroimage*, *16*(3), 765-780.

Vigliocco, G., Vinson, D. P., Druks, J., Barber, H., & Cappa, S. F. (2011). Nouns and verbs in the brain: a review of behavioural, electrophysiological, neuropsychological and imaging studies. *Neuroscience & Biobehavioral Reviews*, *35*(3), 407-426.

Wang, C., Peng, D., Guan, L., & Kuang, P. (1999). The role of surface frequencies, cumulative morpheme frequencies, and semantic transparencies in the processing of compound words. *Acta Psychologica Sinica*, *31*(3), 266-273.

Wang, K. (2011). An electrophysiological investigation of the role of orthography in accessing meaning of Chinese single-character words. *Neurosci Lett*, *487*(3), 297-301. https://doi.org/10.1016/j.neulet.2010.10.041

Wang, K., Mecklinger, A., Hofmann, J., & Weng, X. (2010). From orthography to meaning: an electrophysiological investigation of the role of phonology in accessing meaning of Chinese single-character words. *Neuroscience*, *165*(1), 101-106.

Wei, Y., Niu, Y., Taft, M., & Carreiras, M. (2023). Morphological decomposition in Chinese compound word recognition: Electrophysiological evidence. *Brain and Language*, *241*, 105267.

Wei, Y., Wang, J., Wang, H., & Paz-Alonso, P. M. (2024). Functional interactions underlying visuospatial orthographic processes in Chinese reading. *Cerebral Cortex*, *34*(9), bhae359.

Wong, A.-K., Wu, Y., & Chen, H.-C. (2014). Limited role of phonology in reading Chinese two-character compounds: Evidence from an ERP study. *Neuroscience*, *256*, 342-351.

Wu, C. Y., Ho, M. H., & Chen, S. H. (2012). A meta-analysis of fMRI studies on Chinese orthographic, phonological, and semantic processing. *Neuroimage*, *63*(1), 381-391. https://doi.org/10.1016/j.neuroimage.2012.06.047

Wu, J., Chang, J., Qiu, Y., & Joseph, D. (2020). The temporal process of visual word recognition of Chinese compound: Behavioral and ERP evidences based on homographic morphemes [in Chinese]. *Acta Psychologica Sinica*, *52*(2), 113-127.

Wulff, D. U., De Deyne, S., Jones, M. N., Mata, R., & Aging Lexicon, C. (2019). New Perspectives on the Aging Lexicon. *Trends Cogn Sci*, *23*(8), 686-698. https://doi.org/10.1016/j.tics.2019.05.003

Xia, Q., Wang, L., & Peng, G. (2016). Nouns and verbs in Chinese are processed differently: Evidence from an ERP study on monosyllabic and disyllabic word processing. *Journal of Neurolinguistics*, *40*, 66-78. https://doi.org/10.1016/j.jneuroling.2016.06.002

Xu, X., Kang, C., & Guo, T. (2019). Brain electrophysiological responses to emotion nouns versus emotionless nouns. *Journal of Neurolinguistics*, *49*, 144-154. https://doi.org/10.1016/j.jneuroling.2018.10.003

Yang, H., Lin, Q., Han, Z., Li, H., Song, L., Chen, L., He, Y., & Bi, Y. (2017). Dissociable intrinsic functional networks support noun-object and verb-action processing. *Brain and Language*, *175*, 29-41.

Yang, J., Tan, L. H., & Li, P. (2011). Lexical representation of nouns and verbs in the late bilingual brain. *Journal of Neurolinguistics*, *24*(6), 674-682.

Yang, Y., Liang, D., Gu, J., Weng, X., & Feng, S. (2002). A neurolinguistic study on the classification of nouns and verbs in Chinese [in Chinese]. *Linguistic Sciences*, *1*, 31-46.

Yang, Y., Zhu, Z., & Chen, Q. (2022). Neurolinguistics in China. In *The Palgrave Handbook of Chinese Language Studies* (pp. 1-48). Springer.

Yu, A., Chen, M. S. Y., Cherodath, S., Hung, D. L., Tzeng, O. J. L., & Wu, D. H. (2019). Neuroimaging evidence for sensitivity to orthography-to-phonology conversion in native readers and foreign learners of Chinese. *Journal of Neurolinguistics*, *50*, 53-70. https://doi.org/10.1016/j.jneuroling.2018.07.002

Yu, X., Law, S. P., Han, Z., Zhu, C., & Bi, Y. (2011). Dissociative neural correlates of semantic processing of nouns and verbs in Chinese—A language with minimal inflectional morphology. *Neuroimage*, *58*(3), 912-922.

Yum, Y. N., Law, S.-P., Mo, K. N., Lau, D., Su, I.-F., & Shum, M. S. (2016). Electrophysiological evidence of sublexical phonological access in character processing by L2 Chinese learners of L1 alphabetic scripts. *Cognitive, Affective, & Behavioral Neuroscience*, *16*(2), 339-352.

Yum, Y. N., Law, S. P., Su, I. F., Lau, K. Y., & Mo, K. N. (2014). An ERP study of effects of regularity and consistency in delayed naming and lexicality judgment in a logographic writing system. *Front Psychol*, *5*, 315. https://doi.org/10.3389/fpsyg.2014.00315

Zhang, B., & Peng, D. (1992). Decomposed storage in the Chinese lexicon. In *Advances in Psychology* (Vol. 90, pp. 131-149). Elsevier.

Zhang, H., Su, I.-F., Chen, F., Ng, M. L., Wang, L., & Yan, N. (2020a). The time course of orthographic and semantic activation in Chinese character recognition: evidence from an ERP study. *Language, Cognition and Neuroscience*, *35*(3), 292-309.

Zhang, J., Wu, C., Meng, Y., & Yuan, Z. (2017). Different neural correlates of emotion-label words and emotion-laden words: An ERP study. *Frontiers in human neuroscience*, *11*, 455.

Zhang, J., Wu, C., Yuan, Z., & Meng, Y. (2018). Different early and late processing of emotion-label words and emotion-laden words in a second language: An ERP study. *Second Language Research*, *36*(3), 399-412. https://doi.org/10.1177/0267658318804850

Zhang, J. X., Fang, Z., Du, Y., Kong, L., Zhang, Q., & Xing, Q. (2012a). Centro-parietal N200: An event-related potential component specific to Chinese visual word recognition. *Chinese Science Bulletin*, *57*(13), 1516-1532. https://doi.org/10.1007/s11434-011-4932-y

Zhang, J. X., Xiao, Z., & Weng, X. (2012b). Neural evidence for direct meaning access from orthography in Chinese word reading. *Int J Psychophysiol*, *84*(3), 240-245. https://doi.org/10.1016/j.ijpsycho.2012.02.013

Zhang, Q., Wang, H., Luo, C., Zhang, J., Jin, Z., & Li, L. (2019). The neural basis of semantic cognition in Mandarin Chinese: A combined fMRI and TMS study. *Hum Brain Mapp*, *40*(18), 5412-5423. https://doi.org/10.1002/hbm.24781

Zhang, Q., Zhang, J. X., & Kong, L. (2009). An ERP study on the time course of phonological and semantic activation in Chinese word recognition. *Int J Psychophysiol*, *73*(3), 235-245. https://doi.org/10.1016/j.ijpsycho.2009.04.001

Zhang, W., Yan, Z., Dong, Q., Dong, J., Lu, Y., Duan, X., & Yan, H. (2024). Semantic transparency modulates the semantic perception of morphemes: Evidence from RSA of BOLD signals. *Neuroscience Letters*, *823*, 137649.

Zhang, X., Yang, J., Wang, R., & Li, P. (2020b). A neuroimaging study of semantic representation in first and second languages. *Language, Cognition and Neuroscience*, *35*(10), 1223-1238. https://doi.org/10.1080/23273798.2020.1738509

Zhao, R., Fan, R., Liu, M., Wang, X., & Yang, J. (2017). Rethinking the function of brain regions for reading Chinese characters in a meta-analysis of fMRI studies. *Journal of Neurolinguistics*, *44*, 120-133. https://doi.org/10.1016/j.jneuroling.2017.04.001

Zhou, A., Yin, Y., Zhang, J., & Zhang, R. (2016). Does font type influence the N200 enhancement effect in Chinese word recognition? *Journal of Neurolinguistics*, *39*, 57-68. https://doi.org/10.1016/j.jneuroling.2016.03.001

Zhou, W., Kwok, V. P. Y., Su, M., Luo, J., & Tan, L. H. (2020). Children's neurodevelopment of reading is affected by China's language input system in the information era. *npj Science of Learning*, *5*(1). https://doi.org/10.1038/s41539-020-0062-0

Chapter 3
Morphology and Morphological Processing

3.1 Introduction

Morphology, as the study of word structure and the ways in which smaller linguistic units combine to form words, constitutes a fundamental domain of linguistic inquiry and has become a focal point in psycholinguistics and cognitive neuroscience. As one of the central subfields of linguistics, morphology plays a crucial role in explaining how languages encode meaning, how words are formed and related to each other, and how these processes reflect the organization of the human mental lexicon (Aronoff 1993; Booij 2012). Beyond its theoretical role, morphology serves as a bridge between linguistic description and experimental research, offering a framework to investigate the mechanisms of language comprehension, word recognition, and cognitive processing (Aronoff and Fudeman 2022; Spencer 1991). It provides the theoretical and empirical basis for understanding how words are represented, processed, and learned in the mind, thereby shaping our knowledge of both linguistic competence and actual language performance (Aronoff 1993; Berthiaume et al. 2018b; Booij 2012; Bornkessel-Schlesewsky and Schlesewsky 2009; Taft and Forster 1975). Consequently, morphology is not only central to linguistic theory but also indispensable for broader models of the human cognition.

Within this broad research tradition, Chinese morphology occupies a particularly unique position. Unlike alphabetic languages that exhibit rich inflections, Chinese is typologically characterized as a morpho-syllabic and largely analytic language in which compounding and derivation are the dominant morphological processes (Packard 2000; Pae 2018). Chinese characters often encapsulate both semantic and phonological information, and their visual-spatial layout introduces further complexity to morphological processing. These typological characteristics raise fundamental questions about the extent to which theories of morphological processing derived from alphabetic languages can be generalized. Moreover, because Chinese writing system is logographic in nature, research on Chinese morphology not only

F. Gao, *The When, Where, and How of Morphological Processing in the Human Brain*, https://doi.org/10.1007/978-981-95-5981-7_3

provides language-specific insights but also offers critical evidence for testing the universality of reading theories.

Against this backdrop, this chapter aims to provide a comprehensive review of morphology and its processing, with a particular focus on Chinese. Section 3.2 discusses the definitions of morphology and theoretical perspectives from different traditions. Section 3.3 outlines the typological characteristics of morphology across languages, which sets the stage for examining Chinese. Section 3.4 introduces the definition and classification of Chinese morphology. Section 3.5 turns to the mental lexicon and the role of morphology in lexical representation and processing. Finally, Section 3.6 reviews key research issues and gaps in Chinese morphological processing, including its representational units, neural basis, and multifaceted characteristics. By synthesizing these perspectives, this chapter seeks to situate Chinese morphology within the broader science of reading and to highlight avenues for future research.

3.2 Morphology: Definitions and Theoretical Perspectives

Morphology is traditionally defined as the study of the internal structure of words and the rules governing word formation (e.g., Aronoff and Fudeman 2022; Bauer 2003; Matthews 1991). It concerns how morphemes, the smallest units of meaning, are combined to create complex words and how these units function within a language system (Aronoff and Fudeman 2022; Lieber 2021; Spencer 1991). As one of the core subfields of linguistics, morphology bridges phonology, syntax, and semantics, making it central to understanding how linguistic form encodes meaning. This general definition highlights morphology as both a descriptive and explanatory domain: it describes linguistic patterns and seeks to explain their cognitive and communicative underpinnings (Booij 2012).

From a theoretical perspective, different models have been proposed to account for word structure. The *item-and-arrangement* model views morphemes as discrete building blocks, where words are formed by concatenating morphemic units. In contrast, the *item-and-process* model emphasizes morphological operations, such as affixation or reduplication, that transform base forms into complex words (Hockett 1954). These two perspectives highlight a fundamental tension between morphology as a system of combinatorial pieces versus morphology as a system of rules and processes. Later, the *word-and-paradigm* approach (Matthews 1974) expanded this view by stressing paradigmatic relations among words, situating morphology within the larger grammatical system.

Within generative grammar, morphology has often been treated as part of syntax or as an autonomous component. The *Lexicalist Hypothesis* (Chomsky 1970) argues that word formation rules operate in the lexicon rather than in syntax, a view further developed in Aronoff's theory of word formation (Aronoff 1976). Generative morphology focuses on the rules that predict possible word forms and constrain morphological productivity. More recent perspectives, such as Construction Morphology

(Booij 2010), view morphological patterns as constructions (i.e., systematic form–meaning pairings) that are organized and stored within a structured lexicon. These models illustrate the theoretical diversity in explaining how morphology interacts with other components of grammar.

To summarize, definitions and theoretical perspectives on morphology converge on the view that morphology is an indispensable component of linguistic structure. Whether approached from structuralist, generative, or construction-based frameworks, morphology consistently provides the tools for analyzing how words are internally organized and systematically related to one another (Aronoff 1993; Spencer 1991). These approaches underscore both the independence of morphological structures and their integration within the overall organization of language. By synthesizing these insights, researchers have established morphology as a core subfield of linguistics, laying the groundwork for further inquiry into how specific languages instantiate general morphological principles.

3.3 Typological Characteristics of Morphology

Morphological typology classifies languages according to their structural and functional strategies of word formation, highlighting both universal principles and cross-linguistic variation. Research on cross-linguistic variation has clarified how languages differ in their use of inflection, derivation, and compounding, with these categories serving as essential comparative concepts for describing morphology (Haspelmath 2010; Haspelmath and Sims 2010). Typological perspectives, also enriched by works such as Comrie (1989) and Bybee (1985, 2015), demonstrate that morphology is not only a linguistic subsystem but also a key domain for understanding the diversity of language architecture.

Inflection Inflectional morphology involves grammatical markers that encode categories such as number, tense, aspect, case, and agreement. In particular, inflection does not change the core meaning or the word class of the base word. Instead, it ensures that words fit into the grammatical structures of sentences. For example, in English, the verb "walk" can be inflected for tense (walks, walked, walking), while the noun "cat" can be inflected for number (cats). These processes adjust a word's form to reflect syntactic requirements, but the lexical entry remains the same.

Cross-linguistic typology shows a striking contrast between languages with extensive inflectional systems (e.g., Russian, Turkish, Latin) and languages with (arguably) little or no inflection (e.g., Chinese, Vietnamese). Comrie (1989) observed that the degree of inflection correlates with broader typological properties such as word order and syntactic alignment. For instance, highly inflected languages entail greater word order flexibility because grammatical relations are encoded morphologically, whereas languages with little or no inflection tend to rely more heavily on fixed word order to signal syntactic roles. Haspelmath (2002) further emphasized that while fusional and agglutinative languages encode grammatical relations morphologically, analytic languages rely primarily on syntax and word order. These findings underscore the typological continuum from highly inflected to isolating languages.

Derivation Derivational morphology, by contrast, creates new lexical items through the addition of prefixes, suffixes, or other word-formation strategies (Aronoff and Fudeman 2022; Spencer 1991). It often changes the meaning of the base and may also alter its grammatical category. For instance, *happy* → *unhappy* illustrates a meaning modification without changing grammatical category (adjective → adjective), while *beauty* → *beautiful* (noun → adjective) or *modern* → *modernize* (adjective → verb) involve category-changing derivation. English derivation relies heavily on prefixation and suffixation. Common prefixes include un- (*unfair*), re- (*rewrite*), and dis- (*disrespect*). Common suffixes include -ness (*kindness*), -er (*teacher*), and -ly (*quickly*).

However, while languages such as English and German make extensive use of derivational affixes (*happi**ness**, Freundlich**keit***), other languages, including Chinese, rely much more heavily on compounding for lexical innovation. Bybee (1985) noted that derivation tends to be more semantically transparent than inflection, reflecting its role in enriching lexical categories. This cross-linguistic variation reveals that derivation, while universal in principle, differs substantially in productivity, transparency, and semantic scope.

Compound Compounding, the process of forming new words from independent lexemes, is arguably the most widespread word-formation strategy across languages (Bauer 2001; Lieber and Štekauer 2009). Haspelmath and Sims (2010) and Booij (2010) mentioned that compounds may range from semantically transparent forms (e.g., *toothbrush*) to highly lexicalized and opaque items (e.g., *understand*). While compounding exists in nearly all languages, its dominance varies. Chinese provides a salient example of compounding as the primary mode of word formation, with disyllabic compounds being the most common lexical units (Packard 2000; Tian and Baayen 2022; Zhou et al. 2009). In contrast, in languages like English or German, compounding coexists with derivational and inflectional systems, all of which are considered productive (Booij 2010; Lieber and Štekauer 2009).

In sum, typological classifications of inflection, derivation, and compounding reveal both the shared mechanisms and the unique profiles of individual languages. While fusional and agglutinative languages are often marked by rich inflection, analytic languages such as Chinese show very limited inflection and depend mainly on compounding. These findings, discussed extensively by Haspelmath and Sims (2010) and others, underscore the need to situate Chinese morphology within this broader typological landscape. The following Sect. 3.4 will therefore focus specifically on the definition and classification of Chinese morphology, highlighting how its distinctive characteristics contribute to our understanding of morphological theory and language processing.

3.4 Chinese Morphology: Definition and Classification

3.4.1 The Concept of Chinese Morphology

The term morphology, originating from the Greek word *morphē*, meaning "form", "shape", or "appearance", was traditionally associated with the study of inflectional paradigms in Indo-European languages, emphasizing changes in word form to express grammatical categories. In this sense, early Chinese linguists translated morphology as 形态学/xing2 tai4 xue2/, literally "the study of forms". Yet this definition does not align well with the structure of Chinese, where inflection is largely limited or entirely lacking. Consequently, scholars have argued that for Chinese, morphology should be more appropriately translated as 构词法 /gou4 ci2 fa3/ ("the rules of word formation") or 词法 /ci2 fa3/ ("the grammar of words"), capturing the centrality of compounding and derivation rather than inflection (see the book cover of Packard 2000). This terminological shift reflects the necessity of adapting morphological theory to language-specific typological properties. Accordingly, this book adopts "构词法" as the working definition of Chinese morphology, that is, the study of how Chinese words are formed and how their internal structures are organized. In other words, Chinese morphology does not primarily concern inflectional paradigms, but rather focuses on the principles and mechanisms of word-formation: how morphemes combine through compounding, derivation, reduplication, or other processes to create new lexical items. This approach captures both the external productivity of word formation and the internal organization of word structure, aligning more closely with the typological properties of Chinese.

From a morphological typology perspective, Chinese is typically classified as an isolating language, with little or no inflectional morphology compared to fusional or agglutinative languages (Liao 2014; Xu 2005). For example, English relies heavily on inflections such as *work–works–worked–working* to encode grammatical categories, while also employing derivational processes (*work–worker*). In contrast, Chinese grammar makes limited use of inflection, with controversial cases such as the plural marker -们 /men2/ (e.g., in 人们/ren2 men/, "people") or verbal reduplication (走走/zou3 zou/, *walk-walk,* "walk a bit"). Instead, compounding constitutes the dominant mechanism of word formation, accounting for more than 70% of the lexicon (Tian and Baayen 2022; Zhou et al. 2009). This feature has profound implications: while morphology in Indo-European languages is often understood as changes to word forms, Chinese morphology emphasizes the combinatory principles of morphemes to generate new lexical items.

A long-standing controversy concerns whether Chinese possesses genuine morphological inflection. Some scholars (e.g., Xu 2025) tended to recognize elements such as -们 or reduplication as inflectional markers. In contrast, structuralist approaches (e.g., Huang and Liao 2011) argue that such phenomena do not constitute inflection in the conventional sense, but rather belong to derivation or compounding. For instance, –们 has been reinterpreted as a derivational suffix restricted to human nouns, and reduplicated verbs as compounds formed from identical morphemes. This debate illustrates how theoretical frameworks, originally based on alphabetic languages, struggle to accommodate Chinese linguistic data, and why a reconceptualization of "morphology" is necessary.

Despite these disagreements, consensus has emerged that compounding and derivation form the backbone of Chinese morphology. In order to capture its complexity, Chinese vocabulary can be broadly classified into three categories: (1) compounding (including some reduplications), (2) derivation, (3) monomorphemic words. Among these, compounding constitutes the most productive process, while the other categories, especially monomorphemic words, also reflect important historical and typological features of Chinese. The classification and examples are summarized in Table 3.1.

3.4.2 Chinese Compounds

Compounding (复合) is the most productive and dominant morphological strategy in Chinese, accounting for more than 70% of the lexicon. Compound words (复合词) are formed by combining two or more morphemes into larger lexical units, and their internal structures can be systematically classified. Crucially, the overall word meaning emerges from the semantic and functional combination of constituent

Table 3.1 The classifications and examples of Chinese morphology

Number of morphemes	Category	Sub-category	Examples
Polymorphemic word	Compound	Subordinate	体温 /ti3 wen1/, "body-temperature: temperature"
		Coordinate	助理 /zhu4 li3/, "help-arrange: assistant"
		Verb–Object	看书 /kan4 shu1/, "read-book: read a book"
		Complement	打倒 /da3 dao3/, "strike-down: overthrow"
		Subject–Predicate	地震 /di4 zhen4/, "earth-shake: earthquake"
		Reduplication	妈妈 /ma1 ma/, "mom-mom: mom"
	Derivation	Prefixes	老虎 /lao3 hu3/, "old-tiger, tiger"
		Suffixes	刷子 /shua1 zi/, "brush-ZI: brush"
Monomorphmic word	Monosyllabic Single-Morpheme		山 /shan1/, "mountain"
	Disyllabic Monomorphemic		葡萄 /pu2 tao/, "grape"
	Phonological Reduplication		猩猩 /xing1 xing/, "gorilla"
	Phonetic Loanwords		咖啡 /ka1 fei1/, "coffee"

morphemes. Building on a structuralist perspective (Huang and Liao 2011),[1] Chinese compounds can be further classified into five major subtypes, along with a distinctive category of reduplication.

(1) Subordination (偏正式, "Modifier–Head"). Subordinate compounds consist of a modifier morpheme and a head morpheme, where the modifier restricts, qualifies, or further specifies the meaning of the head. They constitute 54% of the total number of disyllabic compounds (Su 2016). This category can be further divided into two subtypes: Attributive–Head (定中) and Adverbial–Head (状中).

a. Attributive–Head. The first morpheme functions as an adjectival or nominal modifier, while the second morpheme is the head noun. E.g.:

黑板 (/hei1 ban3/, "black-board") → 黑 (black) modifies 板 (board).

红花 (/hong2 hua1/, "red-flower") → 红 (red) modifies 花 (flower).

b. Adverbial–Head. The first morpheme serves as an adverbial modifier that indicates manner, degree, or circumstance, while the second morpheme is usually a verb or an adjective. E.g.:

热销 (/re4 xiao1/, "hot-sell") → 热 (hot) modifies 销 (sell), indicating that the item is selling very well.

火红 (/huo3 hong2/, "fire-red") → 火 (fire) functions metaphorically as a modifier to 红 (red), intensifying the quality of redness.

(2) Coordination (联合式/并列式, "Parallel"). Both morphemes are semantically equal and combine in a coordinate relationship, usually implying "and". E.g.:

花草 (/hua1 cao3/, "flower-grass → plants") → 花 (flower) and 草 (grass) are co-equal.

父母 (/fu4 mu3/, "father-mother → parents") → 父 (father) and 母 (mother) are parallel.

(3) Verb–Object (述宾式/动宾式). The first morpheme is a verb, and the second is its object. The structure mirrors a verb–object phrase. E.g.:

吃饭 (/chi1 fan4/, "eat-food") → 吃 (eat) governs 饭 (food).

看书 (/kan4 shu1/, "read-book") → 看 (read) governs 书 (book).

(4) Complement (补充式). One morpheme serves as the core (verb or noun), while the other functions as a complement, specifying the result, state, or additional meaning of the head morpheme.

a. Verb–Complement (动补式).

The verb precedes, and the complement specifies the result or extent of the action. E.g.:

吃完 (/chi1 wan2/, "eat-finish → finish eating") → 吃 (eat) + 完 (finish).

看懂 (/kan4 dong3/, "see-understand → understand by looking") → 看 (see) + 懂 (understand).

b. Noun–Complement (名补式).

The noun precedes, and the complement qualifies or extends the meaning of the noun, often specifying measure, shape, or subtype. E.g.:

[1] Huang and Liao (2011) use the term *héchéngcí* (合成词, "compound/composite words" in their framework) to refer broadly to polymorphemic words in Chinese, which include compounds (five structural subtypes), derivations, and reduplication forms.

纸张 (/zhi3 zhang1/, “paper-sheet → sheets of paper”) → 纸 (paper) + 张 (sheet, *measure word*).

房间 (/fang2 jian1/, “house-space → room”) → 房 (house) + 间 (space, *measure word*).

(5) Subject–Predicate (主谓式). The first morpheme functions as the subject, and the second as the predicate, forming a sentence-like structure. E.g.:

日出 (/ri4 chu1/, “sun-rise”) → 日 (sun) is the subject, while 出 (rise) is the predicate.

心急 (/xin1 ji2/, “heart-anxious → impatient”) → 心 (heart) as subject, and 急 (anxious) as predicate.

(6) Compounding Reduplication (重叠式). Although some scholars have treated Chinese reduplication as inflectional morphology (Xu 2025), we argue that it is better understood as a special subtype of compounding, since it involves the combination of identical root morphemes into a new lexical unit. However, reduplication in Chinese tends to be more colloquial and pragmatic in nature, and compared with other types of compounds, it is generally less lexicalized. Notably, Huang and Liao (2011) distinguishes between phonological reduplication, which involves the repetition of non-morpheme syllables, and compounding reduplication, which involves the repetition of free morphemes. The present discussion concerns the latter. It can apply to different word classes:

Nouns: 妈妈 (*/ma1 ma/,* “mother-mother”, mother).

Verbs: 走走 (*/zou3 zou/,* “walk-walk”, walk a bit).

Statistical studies show that subordination (54%) and coordination (26%) together dominate, while other categories like verb-resultative (2%) and subject-predicate (1%) are relatively rare (Su 2016). The prevalence of compounding underscores the analytic nature of Chinese, where meaning is built through juxtaposition rather than inflectional morphology.

3.4.3 Chinese Derivations

Derivational morphology (派生) in Chinese, though less productive than compounding, plays a significant role in expanding the lexicon and creating nuanced semantic distinctions. Unlike inflection, which primarily serves grammatical purposes, derivation changes the lexical category or meaning of a word. In Chinese, derivation is primarily achieved through affixation, including prefixes and suffixes. While Chinese derivation is less extensive compared to Indo-European languages, it exhibits systematic patterns in word formation (Cao 2003; Li and Thompson 1989; Packard 2000). For instance, the suffix –子 (zi3) in 刷子 (/shua1 zi/, “brush”) forms nouns, while –化 (hua4) in 工业化 (/gong1 ye4 hua4/, “industrialize”) creates verbs of transformation. Prefixes are also common, such as 老- (lao3) in 老虎 (/lao3 hu3/, “tiger”), which conveys familiarity, or 第- (di4) in 第一 (/di4 yi1/, “first”), which creates ordinal numerals. Overall, dictionary-based surveys estimate that derivations account for about 14–15% of multimorphemic words in Chinese (Cao 2003).

Within affixation, Chinese prefixes and suffixes can be further classified by function and lexical domain. Prefixes typically modify or qualify the meaning of a root morpheme without changing its category. Examples include 老– /lao3–/, marking familiarity or social status (e.g., 老张/lao3 zhang1/, "Old Zhang"); 阿– /a1/, a diminutive or affectionate marker (e.g., 阿姨/a1 yi2/, "aunt"); and 第– /di4–/, used in ordinals (第一/di4-yi1/, "first"). Suffixes are more productive and can be divided into several categories:

1. Nominal suffixes, such as –子/zi3/ in 桌子 (/zhuo1-zi/, "table") and –儿 (r) in 花儿 (/hua1-r/, "flower" with diminutive);
2. Verbal suffixes, such as –化/hua4/ in 现代化 (/xian4 dai4 hua4/, "modernize").

Some linguists also noted infix-like derivations (e.g., –得/de/and 不- /bu4/) where morphemes insert between compounds, though this is rare and debated in terms of categorization (Packard 2000). Together, these affixation processes demonstrate how Chinese derivation, while limited compared to compounding, systematically enriches vocabulary by creating new lexical categories and refining semantic scope.

3.4.4 Monomorphemic Words in Chinese

Monomorphemic words (单纯词) also constitute an important component of Chinese vocabulary. Unlike multimorphemic words that are typically formed through derivation or compounding two or more morphemes, monomorphemic words consist of a single morpheme and represent some of the most fundamental lexical units in the language. Their scope is quite broad, encompassing not only monosyllabic morphemes but also disyllabic single-morpheme words, phonological reduplication forms, phonetic loanwords, and onomatopoeic words (Huang and Liao 2011). This category reflects both the historical depth of the Chinese lexicon and its adaptability to cultural and linguistic contact.

3.4.4.1 Monosyllabic Single-Morpheme Words

Monosyllabic single-morpheme words (单音节单语素词) are the most basic lexical items in Chinese, consisting of a single syllable that corresponds directly to one morpheme. Examples include 山/shan1/ ("mountain"), 水/shui3/ ("water"), and 火/huo3/ ("fire"). Such words are often semantically primitive and carry fundamental concepts, forming the backbone of the lexicon.

From the perspective of morpheme property, they belong to free morphemes (自由语素), meaning that they can function independently as words (e.g., 水, "water") or combine with other morphemes to form larger lexical units (e.g., 海水 /hai3 shui3/, "sea-water"; 喝水/he1 shui3/, "drink-water"). This category is in contrast to bound morphemes (黏着语素) that cannot stand alone (e.g., 足- "foot-" can only appear in compounds like 足球/zu2 qiu2/, "football"). Because free morphemes simultaneously serve as both morphemes and words, they are frequently used in psycholinguistic experiments to investigate the hierarchical representation of the Chinese mental lexicon (e.g., Hao 2015; Peng et al. 1998; Zhang et al. 2009, 2023).

3.4.4.2 Disyllabic Monomorphemic Words

This category consists of disyllabic words that nevertheless function as indivisible monomorphemic units (连绵词, or "binding words"). Their meanings cannot be derived from their constituent syllables, and the whole form is treated as a single morpheme. In this sense, they blur the boundary between form and meaning, since their orthographic segmentation does not map onto morphemic segmentation. Examples include 蝴蝶/hu2 die2/ ("butterfly") and 葡萄/pu2 tao2/ ("grape"). Due to this structural peculiarity, disyllabic in form but monomorphemic in nature, they are frequently employed in psycholinguistic research as baselines against which to compare bimorphemic compounds (e.g., Hsu et al. 2019; Taft and Zhu 1995; Wei et al. 2023). In particular, Wei et al. (2023) demonstrated that disyllabic monomorphemic words are processed more holistically than bimorphemic compounds, with reduced evidence for morphological decomposition in behavioral and neural measures. It suggests that this category provides a valuable testing ground for distinguishing between whole-word and morpheme-based representations in the Chinese mental lexicon.

3.4.4.3 Phonological Reduplication

Unlike compounding reduplication, which involves the repetition of free morphemes to form compounds, phonological reduplication (叠音词) is formed by repeating a syllable that does not independently function as a morpheme (Huang and Liao 2011). The resulting form is still a single-morpheme word, disyllabic in structure but monomorphemic in nature.

Examples:

猩猩/xing1 xing/ ("gorilla") → the repeated syllable 猩 does not exist independently as a morpheme.

姥姥/lao3 lao/ ("grandmother") → both syllables are bound together as a single morpheme with an indivisible meaning.

Such words are classified as monomorphemic despite their disyllabic form, and they occupy a unique position in Chinese morphology, distinct from inflection, derivation, or compounding.

3.4.4.4 Phonetic Loanwords

Chinese has also incorporated numerous monomorphemic items through phonetic borrowing, especially for cultural imports, brand names, or foreign concepts. These phonetic loanwords (音译外来词) are formed by approximating the foreign pronunciation with Chinese syllables. A typical example is 沙发/sha1 fa1/, "sofa", which reflects the foreign pronunciation but is treated as a single lexical unit in Chinese.

This phenomenon overlaps what Lavric et al. (2012) described as "pseudomorphological" words (e.g., corner), in which the constitutes may have form-based associations or meanings in isolation (e.g., "corn" and "-er") but are not semantically related to the holistic meaning of the word. For instance, in 沙发, the character 沙 means "sand" and 发 means "emit," yet together they do not yield the meaning "sofa". Instead, the word is directly tied to its foreign origin and processed monomorphemically.

In other cases, the internal characters of a phonetic loanword may lack any independent semantic content altogether. For example, 咖啡 (/ka1 fei1/, "coffee", from western languages) consists of syllables that function only as phonetic carriers, with no lexical meaning in Chinese. These items thus highlight the tension between orthographic transparency and semantic opacity, illustrating how Chinese accommodates foreign vocabulary while preserving its morpho-syllabic writing system.

3.4.4.5 Onomatopoeic Words

Onomatopoeic words (拟声词) in Chinese imitate natural sounds and are particularly prevalent in colloquial speech, children's language acquisition, poetry, and oral narratives, such as 汪汪 (/wang 1 wang1/, "woof", imitating the sound of a dog barking), 叮咚 (/ding1 dong1/, "ding-dong", imitating the sound of a bell, chime, or dripping water), and 哗啦 (/hua1 la1/, "crash", imitating the sound of a splashing or crashing noise such as falling water or breaking objects). These lexical items exemplify phonological iconicity, where the phonetic form of the word bears a resemblance to its referential meaning. In cognitive linguistics, this is understood as a manifestation of sound symbolism and iconicity, which underscores the non-arbitrary relationship between linguistic form and conceptual content (Dingemanse 2012; Perniss et al. 2010).

While onomatopoeia in Chinese holds strong potential for illuminating iconic links in language processing, systematic cognitive and neurolinguistic research remains minimal. Future studies could employ behavioral tasks, neuroimaging, or developmental methodologies to expand understanding of how Chinese speakers perceive and process these iconic forms, both within Chinese context and across cultures.

In sum, Chinese morphology is best conceived as a system of word-formation (构词法) rather than inflectional morphology. Its classification spans monomorphemic words, derivation, and compounding, with compounding accounting for the vast majority of lexical items. The distinctive feature of Chinese lexicon is the blurred boundary between morphemes and words, whereby many morphemes function autonomously as lexemes. This duality not only reflects the typological profile of Chinese as an analytic language but also provides a unique testing ground for psycholinguistic inquiry into lexical representation and processing.

3.5 Mental Lexicon and Morphological Processing

3.5.1 Morphology in the Mental lexicon

The concept of the mental lexicon was first introduced by Treisman (1960) in the context of selective listening, where she used the term to describe the mental storage of lexical information accessed during speech comprehension. Unlike a traditional dictionary, the lexicon here is not a static list of words but rather a dynamic cognitive construct. The metaphorical nature of the term has always been noted: it refers not to an actual place in the brain but to a psychological model of how words and morphemes are represented, organized, and retrieved in real time. This view highlights the mental lexicon as an interface between linguistic knowledge and cognitive processing.

Later developments expanded and systematized this idea. Aitchison (1987) popularized the term in her book *Words in the Mind: An Introduction to the Mental Lexicon*, portraying the mental lexicon as a flexible and constantly evolving system shaped by acquisition, frequency, and context. Most notably, Levelt (1989) placed the mental lexicon at the core of speech production in his influential model. In Levelt's framework, lexical access involves multiple stages: lemma selection (semantic-syntactic representation), word form encoding (phonological specification), and articulation. Crucially, the lexicon is not passive storage but the driving engine of language production, mediating between conceptual intentions and phonological realization. This shift emphasized that morphology and lexicon are not peripheral but central to the architecture of language use.

From a structural perspective, entries in the mental lexicon are often conceived as multi-layered representations. A lexical entry may include semantic information (meaning), syntactic specifications (syntactic category, e.g., verb, noun), morphological features (e.g., inflection), and phonological/orthographic form (e.g., syllabic or segmental encoding). Importantly, these components can be further divided into the lemma, which encodes the semantic and syntactic properties of a word, and the lexeme, which encodes its morphological, phonological, and orthographic realization (see Fig. 3.1). Psycholinguistic evidence suggests that these layers are interconnected but partially independent, allowing for phenomena such as morphological priming or slip-of-the-tongue states (Schiller 2020). This complex internal architecture enables efficient retrieval and flexibility in processing.

Within this architecture, morphology occupies a pivotal position. Morphemes, as the smallest meaning-bearing units, serve both as building blocks of lexical entries and as links between related words (e.g., happy, unhappy, happiness). Morphological relations organize the lexicon into networks or families, guiding both recognition and production. However, it is important to note that most of the classical elaborations on the *mental lexicon*, rooted in Indo-European languages, has treated morphology primarily as inflection (e.g., tense, agreement, case) and derivation. This emphasis reflects typological biases rather than universal cognitive structures. For Chinese, where inflection is minimal or absent, morphology is largely realized through compounding, which structure the lexicon in fundamentally different ways from Indo-European systems. Studying morphology in the Chinese mental lexicon therefore provides a crucial opportunity to broaden and revise general theories of lexical representation.

Additionally, in the domain of L2 acquisition, the mental lexicon undergoes dynamic restructuring as learners integrate new lexical and morphological information. Jiang (2000) proposed a model in which L2 lexical development proceeds through stages: initially, L2 words are represented as formal shells mapped onto L1 semantic entries; as proficiency improves, morphological and syntactic specifications are gradually integrated. Morphological information, however, is often fragile or underrepresented in L2 lexicons, leading to processing differences between L1 and L2 speakers. This suggests that mastery of morphology is a critical bottleneck in achieving native-like lexical processing. For Chinese L2 learners, challenges may be especially pronounced, as compounding and derivation, not inflection, form the backbone of morphological organization. Understanding how morphology is represented and accessed in bilingual mental lexicon thus provides key insights into both language-specific processing and universal mechanisms of lexical learning.

3.5.2 Morphological Processing: Theories and Issues

Morphological processing refers to the mental operations by which speakers recognize, interpret, and produce morphologically legit words. It highlights the psychological and cognitive dimensions of morphology, situating it not merely as a formal

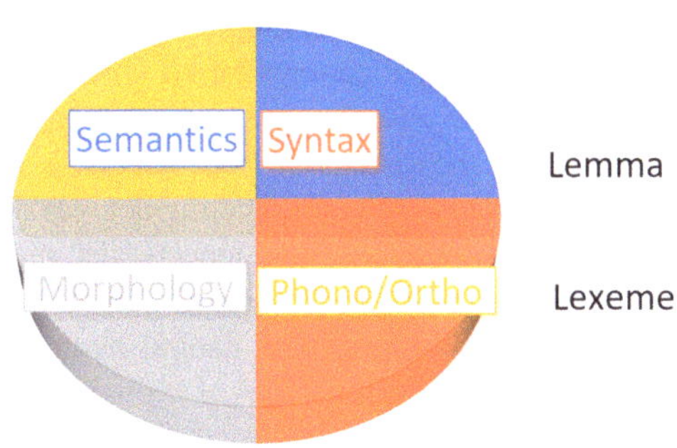

Fig. 3.1 The organization of a lexical entry in the mental lexicon (based on Levelt 1989 and Jiang 2000)

system of word structure but as a dynamic process that unfolds in real time during language comprehension and production. This perspective has become central to psycholinguistics, as it connects linguistic representations with the mechanisms of memory, attention, and learning. Meanwhile, morphological processing is also critical for reading and literacy acquisition, as readers rely on their morphological knowledge to parse and manipulate morphemes to access word meaning efficiently, which underscores its broader cognitive and educational significance (Berthiaume et al. 2018b).

The theoretical landscape of morphological processing has been dominated by two major accounts: the dual-route mechanism and the connectionist approaches. The dual-route mechanism assumes that complex words can be processed either through decomposition into their constituent morphemes or via whole-word storage in the lexicon (Clahsen 1999; Isel et al. 2003; Silva and Clahsen 2008). Regularly inflected forms, such as "walked", are usually parsed through decomposition into stem plus affix, whereas irregular forms like "went" are retrieved directly from memory. This duality allows the model to account for both the rule-governed regularities and the idiosyncratic exceptions that coexist in natural languages, providing a flexible explanation for how speakers process morphological complexity.

Evidence for the dual-route model has been especially robust in Indo-European languages, where inflectional morphology is pronounced. Studies on morphological priming have shown that regularly inflected forms reliably facilitate the access to their stems, consistent with decomposition-based processing (Leminen et al. 2019; Marslen-Wilson et al. 1994). In contrast, irregular forms do not produce the same priming effects, suggesting reliance on direct lexical storage. Neuroimaging findings further supported this distinction, revealing partially dissociable neural pathways for regular and irregular morphology (Allen et al. 2003; Newman et al. 2007). These converging lines of evidence highlight the strength of the dual-route account in capturing the asymmetries of morphological processing, though its explanatory scope remains shaped by typologies dominated by richly inflected Indo-European languages.

In contrast to dual-route models, connectionist approaches conceptualize morphological processing as the emergent outcome of distributed patterns of activation in a neural network, rather than the operation of discrete symbolic rules (Bybee and McClelland 2005; Joanisse and Seidenberg 1999; Kielar et al. 2008). In this framework, both regular and irregular word forms are processed through the same mechanism of associative learning. The network encodes statistical regularities of form-to-meaning mappings based on exposure, allowing it to generalize to novel items while also retaining the idiosyncratic patterns of irregular forms. From this perspective, the distinction between decomposition and whole-word storage is not categorical but gradient, reflecting degrees of similarity and frequency in the input.

Empirical support for connectionist models comes from developmental and cross-linguistic data. For instance, children's overregularization errors (e.g., *goed* for *went*) are interpreted as natural byproducts of the model's learning dynamics rather than as misapplications of explicit rules (Pinker and Ullman 2002). Moreover, frequency effects play a central role, such that high-frequency irregulars are processed more efficiently than low-frequency ones, consistent with predictions of distributed representations (Seidenberg and Gonnerman 2000). Neurocognitive studies further suggested that both regular and irregular forms engage overlapping neural substrates, aligning with the connectionist view of a single mechanism underpinning morphological processing (Joanisse and Seidenberg 1999). While these models well capture gradient effects and frequency sensitivity, they have also been critiqued for underestimating the role of symbolic structure, particularly in languages with productive morphological systems.

As such, the dual-route and connectionist accounts offer contrasting yet complementary perspectives on morphological processing. Dual-route models emphasize symbolic decomposition and rule-based operations, capturing categorical distinctions between regular and irregular forms. By contrast, connectionist models highlight the gradient, frequency-sensitive nature of morphological computation, framing it as the emergent product of distributed learning mechanisms. Despite their differences, both frameworks converge on the idea that morphological processing is shaped by the interplay between stored lexical representations and dynamic processing mechanisms. Importantly, both theories have been primarily developed on the basis of Indo-European languages, especially English and German, where inflection is central. This reliance on inflection-rich systems has raised questions about their generalizability. Languages like Chinese, where compounding and derivation dominate, or Semitic languages with non-concatenative morphology, present challenges that cannot be fully captured by models grounded in inflectional paradigms (Leminen et al. 2019; Schiller 2020). Future research thus needs to expand the empirical scope across typologically diverse languages to refine or possibly reconcile these theoretical accounts.

Morphological processing can be studied across two primary modalities: comprehension and production. Comprehension research investigates how readers and listeners parse and interpret morphologically complex words (e.g., in a lexical decision task), while production studies focus on how speakers assemble morphemes during speech (e.g., in a naming task). Within these modalities, three major morphological types have been the focus of psycholinguistic inquiry: inflection, derivation, and compounding. Inflection has received the most attention, as it is pervasive in Indo-European languages and provides a straightforward testing ground for decomposition versus whole-word storage debates (Golato 2008; Leminen et al. 2019). Derivation has also been widely studied, often in connection with its role in vocabulary development and lexical access. Compounding, however, remains the least investigated, despite its central role in forming new lexical items (Fabregas 2012; Libben 2014; Pollatsek et al. 2011) or in morpho-syllabic languages like Chinese (DeFrancis 1989).

A recurring observation is that research tends to focus more on comprehension over production, inflection over other morphological types, and Indo-European languages over cross-linguistic diversity. This imbalance constrains our understanding of the mechanisms of morphological processing by overgeneralizing from a narrow set of languages and tasks, potentially obscuring modality-specific processes, morphological diversity, and universal versus language-specific principles. Specifically, existing studies on morphological production have largely relied on naturalistic but uncontrolled evidence, such as analyses of speech errors (slips of the tongue) like "*getting your model renosed*" for "*getting your nose remodeled*" (Fromkin 1971; Stemberger 1982) or data from patients with brain lesions (e.g., Schiller et al. 2001). While these approaches provide valuable insights into the structure of morphological representations, they are limited by their reliance on rare, uncontrolled, and often post-hoc data. In contrast to well-controlled psycholinguistic and neurolinguistic experiments, such evidence cannot systematically manipulate variables of interest, establish causal mechanisms, or disentangle competing theoretical accounts of morphological processing. This limitation partly reflects the methodological challenges of collecting production data in this field. Speech production introduces substantial artifacts into EEG signals (e.g., muscle activity contamination) and head-motion artifacts into fMRI, which complicate the analysis of neural processes during real-time production. As a result, production studies are underrepresented compared to comprehension, despite their central role in testing theoretical models of morphological processing. Future research needs to address these gaps by developing innovative paradigms and multimodal methodologies that can capture natural, ecologically valid production processes while minimizing methodological noise.

Further, as Leminen et al. (2019) aptly put it, the research into the three morphological types showed "the good (inflection), the bad (derivation), and the ugly (compounding)" patterns, leaving substantial gaps in our understanding of morphological processing across different morphological types and typological systems. Comparative research is still scarce and has not yet systematically addressed the full range of morphological phenomena, such as reduplication or compounding in isolating languages, which are crucial for testing the universality of existing theories. Expanding the empirical base to understudied languages like Chinese is thus essential for refining and challenging current models.

3.5.3 Morphological Processing and Language Development

Morphology constitutes an essential component of both language processing and vocabulary development. Research has demonstrated its importance in skilled adult reading, children's first language acquisition, adult second language learning, and special populations such as individuals with brain injury or developmental disorders like dyslexia. First, studies of children have consistently shown that sensitivity to morphemes supports decoding, word recognition, and reading comprehension (Carlisle 2000; Kuo and Anderson 2006). Morphological processing provides young readers with analytic strategies for breaking down complex words into meaningful parts, thereby facilitating both vocabulary growth and text comprehension. Importantly, while early literacy skills are grounded in implicit processing, explicit awareness gradually develops during learning, allowing children to consciously manipulate morphemic structures. Evidence from longitudinal studies confirmed that morphological awareness is a strong predictor of later reading achievement, even after controlling for phonological awareness (Nagy et al. 2006).

For L2 learners, morphological knowledge facilitates lexical inference and word recognition, often compensating for limited vocabulary (Jiang 2000; Zhang and Koda 2013). In adults, morphological decomposition enhances lexical access efficiency, while deficits in morphological processing have been observed in dyslexic readers and aphasic patients (Marslen-Wilson and Tyler 2007). Together, these findings highlight morphology as a cross-cutting dimension of linguistic competence across different populations.

In particular, a key conceptual distinction in the literature lies between morphological processing and morphological awareness, terms that are sometimes conflated. Carlisle and Feldman (1995) provided one of the earliest formal definitions of morphological awareness, characterizing it as children's explicit knowledge of the morphemic structure of words and their capacity to reflect upon and manipulate this structure. This definition has since served as a foundational reference in the field, framing morphological awareness as a metalinguistic ability that encompasses conscious analysis of word formation. Later scholars (e.g., Deacon et al. 2013; Guo et al. 2011; Kieffer and Lesaux 2012; Kuo and Anderson 2006) also emphasized that conscious decomposition and analysis of morphemes lie at its core. Morphological awareness thus represents a type of metalinguistic knowledge, that is, the ability to consciously analyze and manipulate the structural features of language beyond its immediate communicative use. Importantly, this capacity is not restricted to children's literacy development but also extends to adult L2 acquisition.

Metalinguistic knowledge encompasses various domains of awareness, each targeting a specific level of linguistic representation (Ke et al. 2023). For instance, *phonological awareness* refers to the capacity to identify and manipulate sound units such as syllables and phonemes, which is foundational to decoding skills in reading. *Orthographic awareness* involves an explicit understanding of the rules, regularities, and permissible patterns of written symbols in a language, supporting accurate word recognition and spelling. Specifically, *morphological awareness* highlights sensitivity to morphemic structure and word-formation processes, enabling learners to decompose complex words into their constituent morphemes or construct new words systematically. These components of metalinguistic knowledge provide a multidimensional framework for explaining how learners gain conscious control over language forms, with significant implications for literacy development and L2 acquisition (Gombert 1992; Koda 2005).

By contrast, morphological processing refers to implicit, automatic computations that occur during online comprehension or production without conscious control. It involves the rapid, unconscious parsing or integration of morphemic information as words are encountered in real time, supporting efficient lexical access and meaning construction. Unlike morphological awareness, which emphasizes explicit reflection and manipulation, morphological processing is measured through time-sensitive tasks such as lexical decision, eye tracking or neuroimaging, which capture the immediate influence of morphemic structure on recognition or production (Bornkessel-Schlesewsky and Schlesewsky 2009; Golato 2008; Leminen et al. 2019). For example, morphological priming studies showed that recognition of a target word (e.g., *teach*) is facilitated by prior exposure to the one with semantically transparent morphological relationship (e.g., *teacher*), indicating that morphemes are automatically activated even when readers are not consciously analyzing structure (Rastle et al. 2004). Neurocognitive evidence further supported this implicit view. ERP components such as the N250 and N400 revealed early sensitivity to morphemic decomposition during word recognition (Morris and Stockall 2012). Thus, morphological processing captures the *online, unconscious mechanisms* that underlie how morphemic information contributes to word identification and language use, complementing the explicit, reflective nature of morphological awareness.

Berthiaume et al. (2018a) proposed a framework distinguishing three related but distinct components concerning "morphological knowledge": specific morphological knowledge, knowledge of rules and regularities, and morphological processing. They argued that it is not the knowledge itself but the procedures through which it is used that can be explicit or implicit. While this clarification is useful, we maintain that tasks and procedures have already largely determined whether morphological knowledge is accessed implicitly or explicitly. For example, lexical decision tasks measure rapid, time-pressured responses and reflect online, implicit processing. In contrast, paper-and-pencil decomposition tasks tap explicit, metalinguistic knowledge. This procedural divide explains why most studies of morphological processing focus on comprehension, while morphological awareness research relies more on production-oriented tasks.

Notably, morphological awareness research has largely concentrated on derivational morphology (see studies reviewed by Berthiaume et al. 2018a), though studies of compounding (especially in Chinese) have gained prominence (Ip et al. 2017; McBride-Chang et al. 2003). Meanwhile, awareness has been extensively studied in children's language development. A consistent finding is that morphological awareness serves as a strong predictor of vocabulary growth and reading comprehension (Carlisle and Feldman 1995; Kuo and Anderson 2006). With literacy instruction, children's capacity to decompose and analyze words improves, underscoring the developmental trajectory from implicit exposure to explicit metalinguistic knowledge.

Studies on morphological processing, on the contrary, go beyond decomposition to also include holistic representation, as postulated in dual-route models that allow both morpheme-based parsing and whole-word access (Marslen-Wilson et al. 1994; Sandra 1990; Zhou and Marslen-Wilson 1995). Evidence from L2 learning further suggested that morphological awareness can scaffold processing efficiency, pointing to a reciprocal relationship between the two domains (Koda 2005, 2015). Developmentally, learners often begin with implicit processing mechanisms, gradually supported and refined by explicit awareness through literacy and instruction. Similarly, in L2 contexts, awareness of morphological structure enables learners to better integrate implicit processing strategies, highlighting the complementary nature of these constructs.

Further, the distinction between implicit processing mechanisms and explicit metalinguistic knowledge suggests that they form a developmental continuum. Morphological processing provides the automatic foundation for comprehension and production, while awareness reflects the explicit mastery of these mechanisms. Research confirmed that morphological awareness is a significant predictor of vocabulary and reading comprehension, partly by strengthening processing efficiency (Bowers and Kirby 2010; Deacon and Kirby 2004). However, interface research bridging processing and awareness is still sparse. Longitudinal studies are especially needed to track how implicit and explicit morphological knowledge interact over time and across populations.

Pedagogically, morphological instruction has been shown to enhance vocabulary learning (Bowers and Kirby 2010; Bowers et al. 2010; Ke and Zhang 2021), yet few studies address how such instruction affects mental representation or neural processing. For example, studies on morphological instruction in classrooms show positive effects on vocabulary expansion and reading comprehension (e.g., Bowers et al. 2010), but do not trace changes at the cognitive or neural level. Future work could explore how tailored instruction targeting both whole-word and morpheme-based strategies might strengthen both awareness and processing, forming a continuum of "morphological instruction → morphological awareness → morphological processing". Such an approach could maximize learning outcomes and connect educational practice with psycholinguistic and neurolinguistic theories.

3.6 Chinese Morphological Processing: Research Issues

Research on Chinese morphological processing occupies a crucial place in psycholinguistics and neurolinguistics, not only because morphology is central to word recognition and lexical organization, but also because Chinese presents typological characteristics that diverge sharply from Indo-European languages. Over the past three decades, a growing body of behavioral and neuroimaging studies has begun to examine how Chinese words are represented and processed in the mental lexicon. Yet, important theoretical and empirical questions remain unresolved. In the following subsections, we summarized the key issues in the field.

3.6.1 Holistic or Decomposed?

Morphological processing in Chinese can be broadly defined as the analysis of internal morphemic information, including both the constituent morphemes and their combinatory relations (Chung et al. 2010). Note that representation and processing are inseparable (Jiang 2000), since how words are stored in the mental lexicon directly shapes how they are accessed in real time. At the center of this debate lies the question of morphological units in the mental lexicon, i.e., whether Chinese words are stored holistically as whole entries, or decomposed into their morphemic components. Classic dual-route mechanisms, developed mainly for Indo-European languages, posit two alternative pathways, including a decomposed route for rule-governed forms and a holistic route for irregular items (Clahsen 1999; Isel et al. 2003; Silva and Clahsen 2008). In the case of Chinese, where morphology is dominated by compounding rather than inflection, the debate has naturally focused on compound words.

The whole-word accounts (Manelis and Tharp 1977) posited that word recognition is achieved through the direct matching of visual stimulus to stored lexical entries. In contrast, the decomposition accounts (Taft and Forster 1975) proposed that multi-morphemic words are first segmented into morphemes before integration. The seminal work from Zhang and Peng (1992) examined the decomposition mechanism in Chinese native speakers' compound word reading by manipulating morphological structures, word and morpheme frequency. The following ERP study (Zhang 1993) demonstrated the N400 sensitivity to Chinese morphemes in target word recognition, thus further supporting the decomposing process at the morpheme level. Hybrid or interactive models (Caramazza et al. 1988; Colé et al. 1997; Fehringer 2012) reconciled these two positions by proposing that both morphemes and whole words are the potential units of representation, with their relative dominance depending on lexical properties such as frequency and semantic transparency. Data from Chinese native speakers appears to support this dual-route mechanism. Behavioral studies using lexical decision and priming paradigms demonstrated that morphemes constitute a necessary level in Chinese lexical access, and it is now accepted that disyllabic compounds in the mental lexicon follow a hybrid representation, where both whole-word and morphemic units are simultaneously engaged in recognition (Hong and Feng 2010). Specifically, high-frequency or semantically opaque compounds often show holistic storage, whereas low-frequency or transparent compounds are processed through decomposition (Wang and Peng 1999).

Word recognition involves a dynamic process of analyzing visual features, accessing lexical entries, and activating semantic representations. An important issue raised in the dual-route framework concerns how constituent morphemes are decomposed and how their form, sound, and meaning are activated at the sub-lexical level. Written Chinese is often described as morpho-syllabic (DeFrancis 1989), with each character corresponding to a syllable that also represents a morpheme. Consequently, morphemic factors at the sub-lexical level are likely to play a critical role in mediating lexical access and meaning retrieval during whole-word reading, making morpheme activation a crucial component of Chinese word recognition.

Both behavioral and neuroimaging studies accumulated evidence for the morphemic effect during Chinese word reading in the human brain. For example, ambiguity and polysemy at the constitute morpheme level would significantly affect the representation of disyllabic compound words. Words with lower sub-lexical ambiguity produced a more negative N400 than those with higher sub-lexical ambiguity when words were presented centrally and to the left visual half-field (VF). In contrast, pseudowords with high sub-lexical ambiguity showed a more negative N400 than low sub-lexical ambiguity with left VF, while low sub-lexical ambiguity pseudowords showed a more negative N400 than high sub-lexical ambiguity pseudowords with right VF presentation (H. W. Huang et al. 2011b). Additionally, the sense facilitation effect at the sub-lexical level was evident in both left and right hemispheres, which was in favor of the assumption of single-entry representation for senses (C. Y. Huang et al. 2011a). These results collectively confirmed the N400 sensitivity to word meaning computation mechanism at the morphemic level.

Besides the semantic modulations, the early stage of morphemic processing was also investigated by using similar materials. Zhao et al. (2017) manipulated the type of target words (with two alternative meanings of the ambiguous morpheme) and the type of prime words (either containing ambiguous morpheme with two meanings or unrelated to the target word) in a visual lexical decision task with masked priming. The ERP results demonstrated an interaction between prime type and target type as reflected by N250 and N400 effects. The modulation of the N250 and N400 components was observed only when the target and prime words shared a morpheme with both the same form and the same meaning. In contrast, only a compromised N250 was observed when prime and target shared the morpheme with distinct meanings. The N250 was therefore interpreted as the morpho-semantic processing at the early stage of morphemic processing.

As such, ERP findings consistently show N400 sensitivity to morphemic semantics, but the precise role of early components (e.g., N250) in form-based decomposition remains debated. Furthermore, neuroimaging has sometimes highlighted whole-word pathways (e.g., left MFG activations, Guo et al. 2022), leaving open the question of whether morphemic decomposition is obligatory or conditional on task demands. These discrepancies underline that while morpheme activation is well established, the degree to which holistic and decomposed pathways are engaged may be modulated by variables such as frequency, transparency, and experimental context.

In summary, research to date indicates that Chinese disyllabic compounds are processed through a hybrid system, where whole-word and morphemic representations co-exist in the mental lexicon. Behavioral studies (e.g., reaction time, priming) provide strong evidence for obligatory morpheme activation, whereas neuroimaging findings suggest that holistic pathways may be more prominent under certain conditions. Yet key questions remain unresolved: if frequency and semantic transparency are carefully controlled, what representational pattern would emerge? How are morphological units represented in the mental lexicon of children or L2 learners, whose experiences with morphemes differ fundamentally from native speakers? These open issues highlight the need for more systematic and developmental research into the representational units of Chinese word recognition.

3.6.2 Neural Representation and Signatures

Extant studies of alphabetic languages, particularly English, have established that morphological processing exhibits distinctive cognitive signatures in the mental lexicon (e.g., Beyersmann et al. 2016). Evidence suggested that inflection, derivation, and compounding may involve partially overlapping but dissociable cortical networks. For instance, research in English highlighted the role of the left IFG and posterior temporal regions in parsing inflectional affixes (Marslen-Wilson and Tyler 2007). Importantly, Leminen et al. (2019) conducted a large-scale synthesis of ERP, MEG, and fMRI studies across languages, and proposed a tripartite account of morphological processing. They showed that inflection is usually processed in a rapid and automatic manner, with early ERP components (e.g., N250) and robust activation in the left fronto-temporal and parahippocampal regions, reflecting efficient rule-based parsing. In contrast, derivation tends to elicit both earlier and later effects in timing and activation in occipito-temporal areas in cortical loci, suggesting higher cognitive demands for semantic integration. Additionally, compounding represents the most complex case, engaging distributed and inconsistent activation patterns across temporal and frontal cortices. This cross-linguistic perspective underscores that while morphological processing is universal in engaging fronto-temporal networks, the timing and degree of specialization differ by various factors including morphological type, task demands, imaging modality, and language typology.

By contrast, the neurobiological basis of Chinese morphology remains poorly understood. Investigating Chinese morphology is crucial not only for elucidating script-specific mechanisms but also for refining universal models of reading (Perfetti et al. 2013; Share 2008). Morphological processing in Chinese necessarily involves complex morpho-semantic computations, since morphemes correspond to single syllable-characters, and the majority of words are disyllabic compounds. The temporal dynamics of Chinese morphological processing have been investigated primarily through studies of disyllabic compound words, especially coordinate compounds (e.g., 花草/hua1 cao3/, flower–grass → "plants"). ERP findings indicate that morphological structure exerts its influence relatively early in the word recognition process. For instance, Chung et al. (2010) reported that a pure morphological structure effect appeared in the 220–300 ms window (P250/P2), followed by the classical N400 reflecting semantic priming and memory network activation. Similarly, Gu et al. (2012) observed that compounds with identical morphological structures elicited stronger P2a responses (150–180 ms) compared to those with mismatched structures, suggesting that readers rapidly compute morphological structure independent of subsequent lexico-semantic analysis. More recent evidence confirmed that semantics may emerge as early as 100–200 ms, with compound recognition shaped more by whole-word semantics than by constituent morpheme meanings, converging toward a *form-and-meaning* account rather than a purely decompositional view (Wang et al. 2024). These results lend support to the view that there exists a specific early-stage ERP signature for morphological structure processing in Chinese compounds. Meanwhile, research has provided complementary evidence that morphological processing may also recruit post-lexical, more conscious mechanisms. For example, words with high morphological productivity, such as subordinate compounds, elicited significantly larger P600 amplitudes than less productive structures like coordinate compounds (Gu and Yang 2010), a pattern interpreted as reflecting late re-analysis of morphological structure. Together, these results suggest that Chinese compound word recognition engages a hybrid time course, which begins with rapid, automatic computation of morphological structure and continues with later, more controlled evaluation when structural or semantic ambiguity arises. This highlights the complexity of Chinese morphological processing, where early sub-lexical cues and later re-analysisboth contribute, and underscores the need for finer-grained designs to clarify how structural and semantic information interact across time.

From a spatial perspective, neuroimaging studies revealed that Chinese morphological processing engages a network overlapping with, yet distinct from, alphabetic languages. Tasks probing morphemic awareness and structural processing consistently activate the left IFG (BA 45/46) and left MTG (BA 21/37), with additional recruitment of the posterior STG (Arredondo et al. 2015; Hsu et al. 2019; Ip et al. 2019; Liu et al. 2009, 2013; Zou et al. 2015). Interestingly, these regions are sensitive to morpho-syllabic characteristics, reflecting the dual role of Chinese characters as both morphemes and syllables. The left MFG, identified as a critical site for the addressed phonology and domain-general processing in Chinese (Tan et al. 2005a, b), may also contribute uniquely to morpho-semantic integration.

Thus, the existing evidence suggests that Chinese morphological processing involves both early sub-lexical operations and later structural re-analysis, engaging a distributed fronto-temporal network. Yet significant gaps remain in clarifying the precise timing ("*when*") and localization ("*where*") of morphological constraints in Chinese word reading. Future work integrating multimodal neuroimaging and cross-linguistic comparisons is needed to determine whether Chinese morphology relies on universal neural systems for morphological processing or recruits language-specific adaptations to its morpho-syllabic script.

3.6.3 Morphological Structure Matters?

Neuroscientific research on morphological processing requires more systematic attention to the role of morphological structure. Leminen et al. (2019) argued that morphological complexity should not be treated as uniform, since polymorphemic words differ in their sub-lexical, lexical, and supra-lexical properties, and individual variation across both items and participants needs to be carefully considered in experimental designs. This perspective highlights the necessity of examining structural variation across different morphological types rather than assuming a homogeneous processing mechanism.

3.6.3.1 Monomorphemic Words, Reduplication, and Derivation

In Chinese, although compounding is the dominant word-formation strategy, other types such as monomorphemic words, derivation, and reduplication are also essential for a comprehensive account of morphological processing. Compared with compounds, these structures are fewer in number but play unique roles in the mental lexicon. Existing studies indicate that monomorphemic words and reduplications are represented differently from compounds in the mental lexicon. Li and Xu (2021), for example, investigated the processing of binding words (e.g., 芙蓉/fu2 rong2/, "lotus") and compounds. They found ERP evidence for the coexistence of expectancy fulfillment and semantic integration, suggesting that binding words require additional semantic composition beyond straightforward morphological processes. Wei et al. (2023) further demonstrated that disyllabic monomorphemic words, though formally similar to compounds, exhibited distinct ERP effects from transparent and opaque compounds at the early stage, thereby supporting the holistic representation of such words. Similarly, Hsu et al. (2019) provided evidence that morphological complexity affects the time course and load of processing, emphasizing the distinctive role of monomorphemic items in Chinese word recognition.

In contrast, reduplication may involve more decompositional mechanisms. Gao et al. (2021) examined Chinese reduplicated verbs and compounding under the condition of Mandarin Tone 3 sandhi, a phenomenon where a low-dipping Tone 3 changes to a rising Tone 2 when it precedes another Tone 3 (e.g., 老鼠, /lao3 shu3/ → /lao2 shu3/, "rat"). The study found that in compound words, both underlying and surface tones are activated and compete during processing, whereas reduplicated words showed stronger reliance on the underlying tonal representation, suggesting they may function more like single-morpheme units. This contrast indicates that reduplication is processed with partial decomposition but also demonstrates unique representational properties distinct from compounds. These findings suggest that Chinese morphological processing is sensitive not only to the presence of multiple morphemes but also to the structural configuration of words.

Neuroimaging studies in Indo-European languages have revealed distinct neural patterns for derivational versus inflectional morphology. Existing evidence tended to show that processing derivational words does not selectively involve the same left lateralized fronto-temporal subsystem that is engaged by the processing of inflected words (Bozic and Marslen-Wilson 2010). Instead, the processing of derivational morphology might engage a larger, bilateral brain network, which underpins whole-word, stem-based lexical access. ERP studies also suggest early morpheme-based detection during visual derivational word recognition. For example, Lavric et al. (2007) compared the neurophysiological responses to masked priming in three conditions: a transparent morphological priming (hunter–HUNT), an opaque morphological priming (corner–CORN), and no morphological relationship (brothel–BROTH). N400 attenuation was observed from morphological priming conditions (both transparent and opaque), relative to non-morphological priming. Similar findings have been reported in Spanish (Havas et al. 2012) and Finnish (Leinonen et al. 2008). Collectively, this evidence underscores that derivational morphology engages early decomposition processes, though with neural profiles that differ from inflectional processing.

To date, no systematic studies have examined the processing of derivational morphology in Chinese. This gap leaves at least two key questions open for future inquiry. First, do Chinese derivational words rely primarily on whole-word representation, morpheme-level decomposition, or a hybrid of both? And how does this compare with the processing routes observed in Indo-European languages? Second, how does the processing of Chinese derivational words align with that of subordinate compounds, given that many Chinese affixes, particularly suffixes like "-人" in "军人" ("soldier"), are historically derived from roots that once served as heads in subordinate structures? Addressing these questions would clarify whether Chinese derivation follows language-general processing principles or reflects language-specific adaptations rooted in its typological profile.

3.6.3.2 Sub-types of Compounding

Among all morphological structures in Chinese, the majority of research has concentrated on compounds. According to Su (2016)'s survey, the five major types of Chinese compound words are distributed unevenly: subordinate structures account for 54%, coordinate for 26%, verb–object for 18%, resultative/complement for 2%, and subject–predicate for 1%. Reflecting this distribution, psycholinguistic and neurolinguistic studies have primarily targeted the contrast between subordinate and coordinate compounds. These two categories not only represent the most frequent patterns in the lexicon but also exhibit distinct cognitive and neural processing profiles, making them the central focus of empirical investigations into Chinese compounding.

Behavioral studies have provided robust evidence for the role of morphological structure in Chinese word processing, especially concerning these two categories. Early work by Zhang and Peng (1992) demonstrated that subordinate and coordinate compounds were processed differently, suggesting that morphological structure influences lexical access. More recently, Liu (2016, 2017) and Liu and McBride-Chang (2010) examined compound word memorization and recognition, showing that subordinate structures were generally easier to memorize and retrieve than coordinate compounds, reflecting their difference in semantic transparency and structural asymmetry. Cui et al. (2018) further observed that for subordinate (modifier–head) compounds, a clear structure priming effect emerged when prime and target shared the same constituent, indicating that constituent activation interacts with structural information in recognition. Yet for coordinate compounds, no such structure priming was observed, suggesting that structural effects are constituent-specific rather than general. Most recently, Hao et al. (2024) extended this line of research by demonstrating that morphological structure interacts with both frequency and semantic transparency, highlighting the interactive nature of Chinese compound word processing. In particular, subordinate and coordinate compounds were recognized more quickly than verb–complement compounds, whereas no significant processing differences were observed among the other structural types. These findings not only reinforce earlier evidence that structural configuration shapes lexical access but also underscore the importance of considering its interaction with other lexical variables in explaining variability in compound word recognition.

Neuroimaging studies have complemented behavioral findings by revealing distinct neural signatures of coordinate and subordinate compounds. Using ERP, Gu and Yang (2010) manipulated the word frequency (high vs. low) and morphological structure productivity (subordinate structure was operationalized as high-productive morphologically, while coordinate structure was considered as low-productive, in light of their corresponding percentage in corpus) and explored their ERP correlates. Words of higher morphological productivity (subordinate structure) elicited significantly larger P600 effect than those of low morphological productivity (coordinate structure), while no early P150 or semantic N400 effect was obtained regarding distinct morphological structures. Meanwhile, Chung et al. (2010) observed a P250 effect for coordinate vs. subordinate compounds, indicating early automatic sensitivity of semantic memory network activation. Hsu et al. (2019), employing MEG, reported more pronounced activations in the left temporal regions when participants processed subordinate compounds relative to coordinate ones. This is consistent with the view that processing subordinate structures involves more complex relational integration. These findings indicate that morphological structure exerts influence at both early and late stages of lexical processing and is reflected in the recruitment of distributed neural networks.

Some studies have looked more closely at the internal conceptual relations of subordinate compounds. Ji and Gagné (2007) demonstrated that conceptual relation information (e.g., cause–effect vs. instrument–object) significantly influences lexical decision times, indicating that relational semantics are activated early during compound recognition. Extending this work, Jia et al. (2013) provided electrophysiological evidence that relation information elicits an N400 effect, confirming that conceptual relations between constituents play an independent role in compound word comprehension. Together, these studies underscore that not only morphological structure, but also the conceptual relation embedded within that structure, modulates compound word processing in Chinese.

Beyond subordinate vs. coordinate distinction, researchers have contrasted processing among other structural subtypes. Gu et al. (2012) compared subordinate and resultative (verb–complement) compounds and found ERP evidence that structural differences are reflected in early P2a responses, suggesting an early sensitivity to structural type. More comprehensively, Hao et al. (2024) compared subordinate, coordinate, verb–object, resultative, and subject–predicate compounds, showing that verb–complement forms exert the greatest processing difficulties, with frequency and semantic transparency interacting with structure. Similarly, Hsu et al. (2019) found that verb–object compounds exhibit distinct neural activation patterns relative to other structures, implicating the left temporal regions in morphological parsing. These findings collectively indicate that the cognitive load of compound recognition varies systematically across structural subtypes.

The structural differences observed across compound types can be interpreted through the lens of spreading activation models. Liu (2017) argued that compound word recognition involves activation spreading from constituent morphemes to the whole-word representation, with efficiency modulated by structural configuration. In subordinate compounds, the modifier and the head establish strong directional links, supporting rapid integration. In contrast, coordinate or verb–complement compounds involve more balanced or sequential activation, which can slow processing or demand additional semantic integration. From this perspective, morphological structure acts as an organizational principle that shapes how morphemic activation spreads within the mental lexicon, explaining why different structures elicit distinct behavioral and neural signatures.

3.6.4 Interplay Between Chinese Morphology and Other Linguistic Components

Chinese is a morpho-syllabic writing system in which each character corresponds to both a syllable and a morpheme (DeFrancis 1989). This unique feature means that morphological information is deeply intertwined with orthographic processing. Evidence from priming and lexical decision tasks shows that orthographic radicals, particularly semantic and phonetic components, facilitate morphemic decomposition (Feldman and Siok 1999; Liu and Xiao 2024). For instance, compound recognition is modulated by the visual-orthographic salience of constituent characters, indicating that orthographic cues support morphological parsing.

Unlike alphabetic languages, Chinese exhibits opaque mappings between orthography and phonology, making morphemes central for access to meaning. Although there is no consensus on whether phonological activation is mandatory in Chinese word recognition, it interacts with morphology in unique ways. In light of Chinese homographic morphemes (同形语素) and ERP techniques, Wu et al. (2020) found that heterophonic-homography (+O-P-S, e.g., 作坊/zuo1 fang/, "workshop"), homophonic-homography (+O+P-S, e.g., 作息/zuo4 xi1/, "work and rest"), and identity (+O+P+S, e.g., 作诗/zuo4 shi1/, "poetry-composing") all produced significant morphological priming effects to the same target word (e.g., 作画/zuo4 hua4/, "painting") as compared to the control. ERP results indicated that early stages of compound word recognition in Chinese involve morpheme-based orthographic processing, where graphemic matching and phonological information dominate, while semantic activation remains uncertain. At later stages, recognition shifts to lemma-based morphological–semantic processing characterized by semantic competition and selection, consistent with the parallel distributed processing model of word recognition (McClelland and Rumelhart 1981; Seidenberg and McClelland 1989).

Morphological structure strongly constrains semantic interpretation. For instance, subordinate compounds (e.g., 黑板/hei1 ban3/, "blackboard") demonstrate stronger semantic compositionality than coordinate ones (e.g., 花草/hua1 cao3/, "plant"), reflecting the role of morpheme-level meaning in compound interpretation (Liu 2017). ERP evidence indicated that semantic ambiguity at the morpheme level modulates N400 amplitude (Huang et al. 2011a, 2011b), suggesting that morphological decomposition provides the semantic building blocks for word meaning. Thus, Chinese morphology plays a mediating role between orthographic form and semantic access.

Finally, morphological structure interacts with syntax, especially in compounding and reduplication. Morphological categories often map onto syntactic roles, such as verb-object compounds (吃饭/chi1 fan4/, "eat-meal") aligning with sentence-level syntax (subject-verb-object). The morphology–syntax interface in Chinese highlights that compounding encodes relations (modifier-head, subject-predicate) that mirror syntactic structures, yet with lexicalized constraints (Gu et al. 2017; Li and Thompson 1989; Packard 2000). Specifically, ERP evidence (Gu and Yang 2010) showed late positivities (P600) associated with the reanalysis of morphological productivity, echoing similar syntactic reanalysis effects. This suggests that morphological and syntactic computations may recruit overlapping neural resources in Chinese. More recently, Yu et al. (2024) compared Mandarin inseparable (e.g., 迟疑/chi2 yi2/, "hesitate") and separable compounds (离合词, e.g., 洗澡/xi3 zao3/, "wash-shower"), as well as compound versus simplex nouns, to investigate whether morphological and syntactic structures elicit distinct neural responses. Results showed that both contrasts triggered a left anterior negativity (LAN) effect with similar scalp distributions, suggesting that morphological and syntactic structure building may rely on shared neurocognitive mechanisms, consistent with non-lexicalist "syntax-all-the-way-down" accounts (e.g., Embick 2015; Harley 2014).

In Chinese, morphological processing interacts dynamically with orthography, phonology, semantics, and syntax. Orthographic cues from radicals and sub-lexical levels support morphemic parsing, while phonological and semantic information modulate early and late stages of compound recognition. Importantly, ERP evidence shows that both morphological and syntactic complexity elicit similar neural signatures, suggesting shared processing mechanisms. Overall, Chinese morphology emerges not in isolation but through continuous interplay with other linguistic components. Future research should further explore how these interactions evolve across developmental stages and bilingual contexts, providing deeper insight into the universality and specificity of morphological processing.

3.6.5 Processing Mechanism in Learning Settings

Research on Chinese morphological processing in learning contexts has highlighted its crucial role in both child language development and L2 acquisition. For Chinese children, morphological awareness is a strong predictor of literacy because of the close morpheme-to-character mapping and the prevalence of compound words (Chung and Hu 2007; Ku and Anderson 2003). Neuroimaging evidence indicated that Chinese–English bilingual children show enhanced activation in the left MTG, reflecting greater sensitivity to semantic information, while also maintaining language-specific pathways in the IFG (Ip et al. 2017). These findings suggest that bilingual exposure fosters both shared and language-specific neural mechanisms of morphological processing.

In L2 learning, Chinese–English bilinguals demonstrate non-selective cross-linguistic activation. ERP studies revealed that bilinguals activate Chinese sub-lexical radicals even when processing English-only tasks, as indicated by early P200 and later N400 effects (Chen and Rossi 2024). This suggests that Chinese morpho-orthographic features extend their influence to English reading. However, transfer effects are asymmetrical: while compounding awareness transfers robustly from Chinese to English, derivational awareness transfers less effectively from English to Chinese (Sun et al. 2023). Such asymmetries highlight the role of structural similarity in shaping cross-linguistic transfer.

Generally, research on Chinese morphological processing in learning settings remains limited, especially for children, L2 learners, and bilinguals. Existing evidence underscores the pivotal role of morphology in literacy development, but its transfer patterns and neural mechanisms are still not fully understood. Future studies should investigate how morphology interacts with other linguistic components in L2 and bilingual contexts, compare bilinguals with diverse language backgrounds, and examine the influence of individual factors such as proficiency, working memory, and cognitive control. Moreover, research should extend to special populations (e.g., children with dyslexia or developmental language disorders) to uncover their neural representations of morphology, which may differ from typically developing learners. Finally, a promising direction is to translate these findings into pedagogy and intervention, providing targeted strategies to strengthen morphological awareness and support literacy acquisition across educational and clinical settings.

3.6.6 *Lexical Variables Related to Morphological Processing*

Lexical variables are important modulators of how Chinese words are recognized and processed. Given the logographic nature of Chinese, variables at the word, morpheme/character, and structural levels jointly shape morphological decomposition and lexical access. The following sections outline key variables that have been shown to influence Chinese morphological processing.

3.6.6.1 Word Properties

Word frequency is one of the most robust predictors of recognition speed and accuracy, with high-frequency words generally processed faster and more accurately than low-frequency ones (Sandra 1994; Sze et al. 2014; Wang and Peng 1999).

Familiarity, often self-rated, reflects subjective exposure and can differ from corpus-based frequency, showing independent contributions to lexical decision performance (Su et al. 2023). For L2 learners in particular, familiarity functions as a direct determinant of word recognition, and it should be distinguished from frequency measures derived from language corpora (Feng 2003).

Word length in Indo-European languages is typically defined by the number of letters. In Chinese, however, since most words are disyllabic, word length is more appropriately quantified by strokes or visual complexity of characters. For example, Jiang et al. (2020) reported a reliable stroke-number effect among L2 learners of Chinese, indicating that visually complex words demand greater orthographic processing resources.

Grammatical category (e.g., noun vs. verb) further modulates processing, with ERP studies revealing N400 modulation across syntactic categories (e.g., Xia et al. 2016). Particularly, category-ambiguous words that can flexibly function as multiple syntactic categories (e.g., 变化/bian4 hua4/ as both "to change" [verb] and "alternations" [noun]), can introduce competition during lexical access, slowing recognition and recruiting additional syntactic and semantic integration processes (Li et al. 2004).

Finally, **concreteness** affects semantic activation, as concrete words evoke stronger imagery and richer semantic networks, leading to faster recognition than abstract ones (Zhang and Zhang 1997).

3.6.6.2 Morpheme Properties

At the morpheme level, **morpheme boundness status** (i.e., whether a morpheme can stand alone as a word or not) shapes how it is processed. Morphemes that can function independently as words are called free morphemes (e.g., "*book*" in English, "人" in Chinese), and they tend to be accessed more directly. By contrast, morphemes that cannot stand alone and must combine with others to form words are known as bound morphemes (e.g., "*re-*" in English, "-子" in Chinese), and their processing relies more heavily on contextual cues.

Morpheme frequency is an important factor in word recognition, with high-frequency morphemes facilitating faster recognition and processing (Cui et al. 2021; Yan et al. 2006). In Chinese, since in most cases (except for disyllabic monomorphemic words and monosyllabic bimorphemic words) a single character corresponds to a morpheme, morpheme frequency largely overlaps with character frequency.

Stroke number refers to the total number of individual brush strokes that make up a Chinese character. As a measure of visual complexity, stroke number might directly affect the perceptual load required for character recognition.

Morphemic neighborhood size, or "family size", denotes the number of different words in which a given morpheme appears as a constituent at the same position. In Chinese, where compounding is highly productive, morphemes with a large family size (e.g., 学/xue2/ "study" appearing in 学习/xue2 xi2/, "study"; 学生/xue2 sheng1/, “student”; 学校/xue2 xiao4/, “school”; 学术/xue2 shu4/, “academic”) provide richer morphological neighborhoods, thereby enhancing lexical access and recognition efficiency. Existing evidence indicated that morphemes embedded in larger families facilitate faster word recognition (particularly for semantically transparent words), reflecting the cumulative activation of morphologically related words (Tsai et al. 2006).

3.6.6.3 Structural Properties

At the structural level, **morphological structure** (e.g., verb-object vs. modifier-noun compounds) determines how internal constituents are parsed, with more canonical structures processed more efficiently (Hao et al. 2024; Liu 2017). **Conceptual structure**, such as thematic vs. attributive relations, modulates compositional processing by constraining how meanings are combined (Ji and Gagné 2007; Jia et al. 2013). **Semantic transparency** is a widely studied variable, where transparent compounds (e.g., 花园/hua1 yuan2/, “flower + garden = garden”) facilitate morphemic decomposition, whereas opaque compounds (e.g., 马虎 /ma3 hu3/,“horse+tiger”, “careless”) rely more on whole-word processing (Wang and Peng 1999). Transparency effects are consistently observed in both behavioral and ERP studies, underscoring its role in balancing morphemic decomposition and holistic lexical access (Han et al. 2025; Lee et al. 2021; Tsang et al. 2022; Zhang et al. 2024).

3.7 Summary

This chapter has provided an overview of morphology and morphological processing, with a particular focus on the typological characteristics of Chinese. It first outlined theoretical perspectives on morphology and the organization of the mental lexicon, then examined key debates regarding whether words are processed holistically or decomposed into morphemic constituents. Special attention was given to Chinese compounding, derivation, and monomorphemic words, as well as the interplay between morphology and other linguistic components, neural signatures, and learning settings. The review also highlighted the influence of lexical variables, including word-, morpheme-, and structure-level properties, on Chinese morphological processing and reading processes.

Compared with Indo-European languages, and especially English, research on Chinese morphology remains relatively underdeveloped. While the study of morphology in alphabetic systems has generated extensive behavioral and neurocognitive evidence, studies of Chinese are fewer in number, narrower in scope, and often restricted to limited populations such as adult native speakers. Much remains to be explored with respect to diverse participant groups (e.g., children, bilinguals, and special populations), a wider range of lexical and contextual variables, and a fuller typological range of morphological structures. Addressing these gaps is crucial, as Chinese morphology, given its logographic writing system and high reliance on compounding, offers unique insights into universal and language-specific mechanisms of word recognition and literacy development.

In sum, this chapter underscores the importance of expanding empirical research on Chinese morphology in order to enrich our understanding of reading science more broadly. To move this field forward, future work must not only broaden its evidence base but also refine its methodological toolkit. The next chapter therefore turns to a review of research methods in the study of Chinese reading and morphological processing, outlining experimental paradigms, neuroimaging techniques, and analytic approaches that can provide the evidence needed to test theoretical claims and address the open questions identified here.

References

Aitchison, J. (1987). *Words in the mind: An introduction to the mental lexicon*. Basil Blackwell.

Allen, M., Badecker, W., & Osterhout, L. (2003). Morphological analysis in sentence processing: An ERP study. *Language and Cognitive Processes*, *18*(4), 405-430.

Aronoff, M. (1976). *Word Formation in Generative Grammar*. MIT Press.

Aronoff, M. (1993). *Morphology by itself: Stems and inflectional classes* (Vol. 22). MIT press.

Aronoff, M., & Fudeman, K. (2022). *What is morphology?* John Wiley & Sons.

Arredondo, M. M., Ip, K. I., Shih Ju Hsu, L., Tardif, T., & Kovelman, I. (2015). Brain bases of morphological processing in young children. *Hum Brain Mapp*, *36*(8), 2890-2900. https://doi.org/10.1002/hbm.22815

Bauer, L. (2001). *Morphological Productivity*. Cambridge University Press.

Bauer, L. (2003). *Introducing linguistic morphology*. Edinburgh university press.

Berthiaume, R., Bourcier, A., & Daigle, D. (2018a). Morphological processing tasks and measurement issues. In *Morphological processing and literacy development* (pp. 48-87). Routledge.

Berthiaume, R., Daigle, D., & Desrochers, A. (2018b). *Morphological processing and literacy development: Current issues and research*. Routledge.

Beyersmann, E., Ziegler, J. C., Castles, A., Coltheart, M., Kezilas, Y., & Grainger, J. (2016). Morpho-orthographic segmentation without semantics. *Psychon Bull Rev*, *23*(2), 533-539. https://doi.org/10.3758/s13423-015-0927-z

Booij, G. (2010). Construction morphology. *Language and Linguistics Compass*, *4*(7), 543-555.

Booij, G. (2012). *The grammar of words: An introduction to linguistic morphology*. Oxford University Press.

Bornkessel-Schlesewsky, I., & Schlesewsky, M. (2009). *Processing syntax and morphology: A neurocognitive perspective* (Vol. 6). OUP Oxford.

Bowers, P. N., & Kirby, J. R. (2010). Effects of morphological instruction on vocabulary acquisition. *Reading and Writing*, *23*(5), 515-537.

Bowers, P. N., Kirby, J. R., & Deacon, S. H. (2010). The effects of morphological instruction on literacy skills: A systematic review of the literature. *Review of educational research*, *80*(2), 144-179.

Bozic, M., & Marslen-Wilson, W. (2010). Neurocognitive contexts for morphological complexity: Dissociating inflection and derivation. *Language and Linguistics Compass*, *4*(11), 1063-1073.

Bybee, J. (1985). *Morphology: A Study of the Relation between Meaning and Form*. John Benjamins.

Bybee, J. (2015). *Language Change*. Cambridge University Press.

Bybee, J., & McClelland, J. L. (2005). Alternatives to the combinatorial paradigm of linguistic theory based on domain general principles of human cognition. *Linguistic Review*, 22, 381-410.

Cao, W. (2003). Investigating the differences between colloquial and literary words in modern Chinese [in Chinese]. *Yuyan jiaoxue yu yanjiu, 6*.

Caramazza, A., Laudanna, A., & Romani, C. (1988). Lexical access and inflectional morphology. *Cognition*, *28*(3), 297-332.

Carlisle, J. F. (2000). Awareness of the structure and meaning of morphologically complex words: Impact on reading. *Reading and Writing*, *12*(3), 169-190.

Carlisle, J. F., & Feldman, L. (1995). Morphological awareness and early reading achievement. *Morphological aspects of language processing, 189209*.

Chen, Y., & Rossi, E. (2024). Sub-Lexical Processing of Chinese–English Bilinguals: An ERP Analysis. *Brain Sciences*, *14*(9), 923.

Chomsky, N. (1970). Remarks on nominalization. In R. Jacobs & P. Rosenbaum (Eds.), *Readings in English Transformational Grammar* (pp. 184–221). Ginn-Blaisdell.

Chung, K. K., Tong, X., Liu, P. D., McBride-Chang, C., & Meng, X. (2010). The processing of morphological structure information in Chinese coordinative compounds: an event-related potential study. *Brain Res*, *1352*, 157-166. https://doi.org/10.1016/j.brainres.2010.06.069

Chung, W.-L., & Hu, C.-F. (2007). Morphological awareness and learning to read Chinese. *Reading and Writing*, *20*(5), 441-461.

Clahsen, H. (1999). Lexical entries and rules of language: A multidisciplinary study of German inflection. *Behavioral and brain sciences*, *22*(6), 991-1013.

Colé, P., Segui, J., & Taft, M. (1997). Words and morphemes as units for lexical access. *Journal of Memory and Language*, *37*(3), 312-330.

Comrie, B. (1989). *Language Universals and Linguistic Typology*. University of Chicago Press.

Cui, L., Cong, F., Wang, J., Zhang, W., Zheng, Y., & Hyona, J. (2018). Effects of Grammatical Structure of Compound Words on Word Recognition in Chinese. *Front Psychol*, *9*, 258. https://doi.org/10.3389/fpsyg.2018.00258

Cui, L., Wang, J., Zhang, Y., Cong, F., Zhang, W., & Hyönä, J. (2021). Compound word frequency modifies the effect of character frequency in reading Chinese. *Quarterly Journal of Experimental Psychology*, *74*(4), 610-633.

Deacon, S. H., Benere, J., & Pasquarella, A. (2013). Reciprocal relationship: children's morphological awareness and their reading accuracy across grades 2 to 3. *Developmental Psychology*, *49*(6), 1113.

Deacon, S. H., & Kirby, J. R. (2004). Morphological awareness: Just "more phonological"? The roles of morphological and phonological awareness in reading development. *Applied Psycholinguistics*, *25*(2), 223-238.

DeFrancis, J. (1989). *Visible speech: The diverse oneness of writing systems*. University of Hawaii Press.

Dingemanse, M. (2012). Advances in the cross-linguistic study of ideophones. *Language and Linguistics Compass*, *6*(10), 654-672.

Embick, D. (2015). *The morpheme: A theoretical introduction* (Vol. 31). Walter de Gruyter GmbH & Co KG.

Fabregas, A. (2012). *Morphology: From data to theories*. Edinburgh University Press.

Fehringer, C. (2012). The lexical representation of compound words in English: evidence from aphasia. *Language Sciences*, *34*(1), 65-75.

Feldman, L. B., & Siok, W. W. (1999). Semantic radicals contribute to the visual identification of Chinese characters. *Journal of Memory and Language*, *40*(4), 559-576.

Feng, L. (2003). An Analysis of Chinese Word Recognition Patterns among Intermediate-Level Learners of Chinese as a Foreign Language [in Chinese]. *Journal of College of Chinese Language and Culture of Jinan University* (03), 26-31+78. https://doi.org/10.16131/j.cnki.cn44-1669/g4.2003.03.005

Fromkin, V. A. (1971). The non-anomalous nature of anomalous utterances. *Language*, 27-52.

Gao, F., Lyu, S., & Lin, C. C. (2021). Processing Mandarin Tone 3 Sandhi at the Morphosyntactic Interface: Reduplication and Lexical Compounds. *Front Psychol*, *12*, 713665. https://doi.org/10.3389/fpsyg.2021.713665

Golato, P. (2008). *Processing French: A psycholinguistic perspective*. Yale University Press.

Gombert, J. E. (1992). *Metalinguistic development*. University of Chicago Press.

Gu, J., Gao, P., Li, S., Yang, J., & Chen, S. (2017). Compounding: Derived from Morphology or Syntax? — An ERP Study on Compound Word Processing [in Chinese]. *Foreign Languages Research*, *34*(05), 36-42. https://doi.org/10.13978/j.cnki.wyyj.2017.05.008

Gu, J., & Yang, Y. (2010). A neuro-electrophysiological study on productivity of Chinese compounding [in Chinese]. *Yuyan Wenzi Yingyong* (3), 98-107.

Gu, J., Yu, L., & Ma, P. (2012). P2a reflects morphology structural processes in brain: Evidence from ERPs [in Chinese]. *Journal of Southwest University (Natural Science Edition)*, *34*(2), 137-144.

Guo, W., Geng, S., Cao, M., & Feng, J. (2022). The brain connectome for Chinese reading. *Neuroscience Bulletin*, *38*(9), 1097-1113.

Guo, Y., Roehrig, A. D., & Williams, R. S. (2011). The relation of morphological awareness and syntactic awareness to adults' reading comprehension: Is vocabulary knowledge a mediating variable? *Journal of literacy research*, *43*(2), 159-183.

Han, Y.-J., Huang, S.-C., & Cheng, S.-K. (2025). ERP Evidence for the Modulation of Semantic Transparency on the Recollection of Two-Character Chinese Words. In *Cognitive and Neural Foundations of Chinese Reading: From Learning to Advanced Processing and Beyond* (pp. 175-197). Springer.

Hao, M. (2015). *The Development of Morpheme Awareness in Chinese Children* [in Chinese]. Beijing Language and Culture Press.

Hao, Y., Wu, C., & Duan, X. (2024). Processing mandarin Chinese compound words by native speakers and second language learners: word frequency, semantic transparency, and word structure. *Sage Open*, *14*(2), 21582440241256249.

Harley, H. (2014). On the identity of roots. *Theoretical linguistics*, *40*(3-4), 225-276.

Haspelmath, M. (2002). *Understanding Morphology*. Arnold.

Haspelmath, M. (2010). Comparative concepts and descriptive categories in crosslinguistic studies. *Language*, *86*(3), 663-687.

Haspelmath, M., & Sims, A. D. (2010). *Understanding Morphology*. Routledge.

Havas, V., Rodriguez-Fornells, A., & Clahsen, H. (2012). Brain potentials for derivational morphology: an ERP study of deadjectival nominalizations in Spanish. *Brain Lang*, *120*(3), 332-344. https://doi.org/10.1016/j.bandl.2011.10.008

Hockett, C. F. (1954). Two models of grammatical description. *Word*, *10*(2-3), 210-234.

Hong, W., & Feng, C. (2010). A Study of the Differences between CSL Learners and Chinese Native Speakers in Chinese Two-Character Compound Word Recognition [in Chinese]. *Modern Foreign Languages*, *33*(4), 387–394.

Hsu, C.-H., Pylkkänen, L., & Lee, C.-Y. (2019). Effects of morphological complexity in left temporal cortex: An MEG study of reading Chinese disyllabic words. *Journal of Neurolinguistics*, *49*, 168-177. https://doi.org/10.1016/j.jneuroling.2018.06.004

Huang, B., & Liao, X. (2011). *Xiandai Hanyu* [in Chinese] (5 ed.). Higher Education Press.

Huang, C. Y., Lee, C. Y., Huang, H. W., & Chou, C. J. (2011a). Number of sense effects of Chinese disyllabic compounds in the two hemispheres. *Brain Lang*, *119*(2), 99-109. https://doi.org/10.1016/j.bandl.2011.04.005

Huang, H. W., Lee, C. Y., Tsai, J. L., & Tzeng, O. J. (2011b). Sublexical ambiguity effect in reading Chinese disyllabic compounds. *Brain Lang*, *117*(2), 77-87. https://doi.org/10.1016/j.bandl.2011.01.003

Ip, K. I., Hsu, L. S., Arredondo, M. M., Tardif, T., & Kovelman, I. (2017). Brain bases of morphological processing in Chinese-English bilingual children. *Dev Sci*, *20*(5). https://doi.org/10.1111/desc.12449

Ip, K. I., Marks, R. A., Hsu, L. S., Desai, N., Kuan, J. L., Tardif, T., & Kovelman, L. (2019). Morphological processing in Chinese engages left temporal regions. *Brain Lang*, *199*, 104696. https://doi.org/10.1016/j.bandl.2019.104696

Isel, F., Gunter, T. C., & Friederici, A. D. (2003). Prosody-assisted head-driven access to spoken German compounds. *Journal of Experimental Psychology: Learning, Memory, and Cognition*, *29*(2), 277.

Ji, H., & Gagné, C. L. (2007). Lexical and relational influences on the processing of Chinese modifier-noun compounds. *The Mental Lexicon*, *2*(3), 387-417.

Jia, X., Wang, S., Zhang, B., & Zhang, J. X. (2013). Electrophysiological evidence for relation information activation in Chinese compound word comprehension. *Neuropsychologia*, *51*(7), 1296-1301. https://doi.org/10.1016/j.neuropsychologia.2013.03.024

Jiang, N. (2000). Lexical representation and development in a second language. *Applied Linguistics*, *21*(1), 47-77.

Jiang, N. A. N., Hou, F., & Jiang, X. I. N. (2020). Analytic Versus Holistic Recognition of Chinese Words Among L2 Learners. *The Modern Language Journal*, *104*(3), 567-580. https://doi.org/10.1111/modl.12662

Joanisse, M. F., & Seidenberg, M. S. (1999). Impairments in verb morphology after brain injury: A connectionist model. *Proceedings of the National Academy of Sciences*, *96* (13), 7592-7597.

Ke, S., & Zhang, D. (2021). Morphological instruction and reading development in young L2 readers: A scoping review of causal relationships. *Studies in Second Language Learning and Teaching*, *11*(3), 331-350.

Ke, S. E., Zhang, D., & Koda, K. (2023). *Metalinguistic awareness in second language reading development*. Cambridge University Press.

Kieffer, M. J., & Lesaux, N. K. (2012). Development of morphological awareness and vocabulary knowledge in Spanish-speaking language minority learners: A parallel process latent growth curve model. *Applied Psycholinguistics*, *33*(1), 23-54.

Kielar, A., Joanisse, M. F., & Hare, M. L. (2008). Priming English past tense verbs: Rules or statistics? *Journal of Memory and Language*, *58*(2), 327-346.

Koda, K. (2005). *Insights into second language reading: A cross-linguistic approach*. Cambridge University Press.

Koda, K. (2015). Development of word recognition in a second language. In *Reading in a second language* (pp. 70-98). Routledge.

Ku, Y.-M., & Anderson, R. C. (2003). Development of morphological awareness in Chinese and English. *Reading and Writing*, *16*(5), 399-422.

Kuo, L.-J., & Anderson, R. C. (2006). Morphological awareness and learning to read: A cross-language perspective. *Educational psychologist*, *41*(3), 161-180.

Lavric, A., Clapp, A., & Rastle, K. (2007). ERP evidence of morphological analysis from orthography: A masked priming study. *Journal of Cognitive Neuroscience*, *19*(5), 866-877.

Lavric, A., Elchlepp, H., & Rastle, K. (2012). Tracking hierarchical processing in morphological decomposition with brain potentials. *Journal of Experimental Psychology: Human perception and performance*, *38*(4), 811.

Lee, H.-J., Cheng, S.-K., Lee, C.-Y., & Kuo, W.-J. (2021). The neural basis of compound word processing revealed by varying semantic transparency and morphemic neighborhood size. *Brain and Language*, *221*, 104985.

Leinonen, A., Brattico, P., Jarvenpaa, M., & Krause, C. M. (2008). Event-related potential (ERP) responses to violations of inflectional and derivational rules of Finnish. *Brain Res*, *1218*, 181-193. https://doi.org/10.1016/j.brainres.2008.04.049

Leminen, A., Smolka, E., Dunabeitia, J. A., & Pliatsikas, C. (2019). Morphological processing in the brain: The good (inflection), the bad (derivation) and the ugly (compounding). *Cortex*, *116*, 4-44. https://doi.org/10.1016/j.cortex.2018.08.016

Levelt, W. J. (1989). *Speaking: From intention to articulation*. MIT Press.

Li, C. N., & Thompson, S. A. (1989). *Mandarin Chinese: A functional reference grammar*. University of California Press.

Li, F., & Xu, X. (2021). Electrophysiological evidence for the coexistence of expectancy fulfillment and semantic integration during the processing of binding and compound nouns. *International Journal of Psychophysiology, 166,* 25-37.
Li, P., Jin, Z., & Tan, L. H. (2004). Neural representations of nouns and verbs in Chinese: an fMRI study. *Neuroimage*, *21*(4), 1533-1541.
Liao, W. R. (2014). Morphology. In C.-T. Huang, Y.-H. Li, & A. Simpson (Eds.), *The Handbook of Chinese Linguistics*. John Wiley & Sons, Inc.
Libben, G. (2014). The nature of compounds: A psychocentric perspective. *Cognitive neuropsychology*, *31*(1-2), 8-25.
Lieber, R. (2021). *Introducing morphology*. Cambridge University Press.
Lieber, R., & Štekauer, P. (2009). *The Oxford Handbook of Compounding*. Oxford University Press.
Liu, D. (2016). The involvement of morphological information in the memorization of Chinese compound words: Evidence from memory errors. *Journal of psycholinguistic research*, *45*(1), 157-176.
Liu, D. (2017). The influence of morphological structure information on the memorization of Chinese compound words. *Reading and Writing*, *30*(8), 1813-1834.
Liu, D., & McBride-Chang, C. (2010). Morphological processing of Chinese compounds from a grammatical view. *Applied Psycholinguistics*, *31*(4), 605-617. https://doi.org/10.1017/s0142716410000159
Liu, L., Deng, X., Peng, D., Cao, F., Ding, G., Jin, Z., Zeng, Y., Li, K., Zhu, L., & Fan, N. (2009). Modality-and task-specific brain regions involved in Chinese lexical processing. *Journal of Cognitive Neuroscience*, *21*(8), 1473-1487.
Liu, L., Tao, R., Wang, W., You, W., Peng, D., & Booth, J. R. (2013). Chinese dyslexics show neural differences in morphological processing. *Dev Cogn Neurosci*, *6*, 40-50. https://doi.org/10.1016/j.dcn.2013.06.004
Liu, Y., & Xiao, F. (2024). Effects of Lexical Properties in L2 Chinese Compound Processing: A Multivariate Approach. *Journal of psycholinguistic research*, *53*(4), 49.
Manelis, L., & Tharp, D. A. (1977). The processing of affixed words. *Memory & Cognition*, *5*(6), 690-695.
Marslen-Wilson, W., Tyler, L. K., Waksler, R., & Older, L. (1994). Morphology and meaning in the English mental lexicon. *Psychological review*, *101*(1), 3.
Marslen-Wilson, W. D., & Tyler, L. K. (2007). Morphology, language and the brain: the decompositional substrate for language comprehension. *Philosophical Transactions of the Royal Society B: Biological Sciences*, *362*(1481), 823-836.
Matthews, P. H. (1991). *Morphology*. Cambridge university press.
Matthews, P. M. (1974). *Morphology: An Introduction to the Theory of Word-Structure*. Cambridge University Press.
McBride-Chang, C., Shu, H., Zhou, A., Wat, C. P., & Wagner, R. K. (2003). Morphological awareness uniquely predicts young children's Chinese character recognition. *Journal of Educational Psychology*, *95*(4), 743-751. https://doi.org/10.1037/0022-0663.95.4.743
McClelland, J. L., & Rumelhart, D. E. (1981). An interactive activation model of context effects in letter perception: I. An account of basic findings. *Psychological review*, *88*(5), 375.
Morris, J., & Stockall, L. (2012). Early, equivalent ERP masked priming effects for regular and irregular morphology. *Brain and Language*, *123*(2), 81-93.
Nagy, W., Berninger, V. W., & Abbott, R. D. (2006). Contributions of morphology beyond phonology to literacy outcomes of upper elementary and middle-school students. *Journal of Educational Psychology*, *98*(1), 134.
Newman, A. J., Ullman, M. T., Pancheva, R., Waligura, D. L., & Neville, H. J. (2007). An ERP study of regular and irregular English past tense inflection. *Neuroimage*, *34*(1), 435-445.
Packard, J. L. (2000). *The morphology of Chinese: A linguistic and cognitive approach*. Cambridge University Press.
Pae, H. K. (2018). *Writing Systems, Reading Processes, and Cross-Linguistic Influences*. John Benjamins Publishing Company.

Peng, D., Liu, Y., & Wang, C. (1998). How is the access representation organized? The relation of polymorphemic word and their morphemes: A Chinese study. *Cognitive Analysis of Chinese Script*, 65-89.

Perfetti, C., Cao, F., & Booth, J. (2013). Specialization and Universals in the Development of Reading Skill: How Chinese Research Informs a Universal Science of Reading. *Sci Stud Read*, *17*(1), 5-21. https://doi.org/10.1080/10888438.2012.689786

Perniss, P., Thompson, R. L., & Vigliocco, G. (2010). Iconicity as a general property of language: evidence from spoken and signed languages. *Frontiers in psychology*, *1*, 227.

Pinker, S., & Ullman, M. T. (2002). The past and future of the past tense. *Trends in cognitive sciences*, *6*(11), 456-463.

Pollatsek, A., Bertram, R., & Hyönä, J. (2011). Processing novel and lexicalised Finnish compound words. *Journal of Cognitive Psychology*, *23*(7), 795-810.

Rastle, K., Davis, M. H., & New, B. (2004). The broth in my brother's brothel: Morpho-orthographic segmentation in visual word recognition. *Psychonomic Bulletin & Review*, *11*(6), 1090-1098.

Sandra, D. (1990). On the representation and processing of compound words: Automatic access to constituent morphemes does not occur. *The Quarterly Journal of Experimental Psychology Section A*, *42*(3), 529-567.

Sandra, D. (1994). The morphology of the mental lexicon: Internal word structure viewed from a psycholinguistic perspective. *Language and Cognitive Processes*, *9*(3), 227-269.

Schiller, N. O. (2020). Neurolinguistic Approaches in Morphology. In *Oxford Research Encyclopedia of Linguistics*. https://doi.org/10.1093/acrefore/9780199384655.013.601

Schiller, N. O., Greenhall, J. A., Shelton, J. R., & Caramazza, A. (2001). Serial order effects in spelling errors: Evidence from two dysgraphic patients. *Neurocase*, *7*(1), 1-14.

Seidenberg, M. S., & Gonnerman, L. M. (2000). Explaining derivational morphology as the convergence of codes. *Trends in cognitive sciences*, *4*(9), 353-361.

Seidenberg, M. S., & McClelland, J. L. (1989). A distributed, developmental model of word recognition and naming. *Psychological review*, *96*(4), 523.

Share, D. L. (2008). On the Anglocentricities of current reading research and practice: the perils of overreliance on an "outlier" orthography. *Psychol Bull*, *134*(4), 584-615. https://doi.org/10.1037/0033-2909.134.4.584

Silva, R., & Clahsen, H. (2008). Morphologically complex words in L1 and L2 processing: Evidence from masked priming experiments in English. *Bilingualism: Language and Cognition*, *11*(2), 245-260.

Spencer, A. (1991). *Morphological theory: An introduction to word structure in generative grammar*. Blackwell.

Stemberger, J. P. (1982). *The lexicon in a model of language production*. University of California, San Diego.

Su, B. (2016). The implicity, diversity and cognitive features of Chinese compounding [in Chinese]. *Xueshu Yanjiu* (1), 162-165.

Su, Y., Li, Y., & Li, H. (2023). Familiarity ratings for 24,325 simplified Chinese words. *Behavior research methods*, *55*(3), 1496-1509.

Sun, X., Marks, R. A., Zhang, K., Yu, C. L., Eggleston, R. L., Nickerson, N., Chou, T. L., Hu, X. S., Tardif, T., & Satterfield, T. (2023). Brain bases of English morphological processing: A comparison between Chinese-English, Spanish-English bilingual, and English monolingual children. *Developmental science*, *26*(1), e13251.

Sze, W. P., Rickard Liow, S. J., & Yap, M. J. (2014). The Chinese Lexicon Project: a repository of lexical decision behavioral responses for 2,500 Chinese characters. *Behav Res Methods*, *46*(1), 263-273. https://doi.org/10.3758/s13428-013-0355-9

Taft, M., & Forster, K. I. (1975). Lexical storage and retrieval of prefixed words. *Journal of verbal learning and verbal behavior*, *14*(6), 638-647.

Taft, M., & Zhu, X. (1995). The representation of bound morphemes in the lexicon: A Chinese study. *Morphological aspects of language processing*, 293-316.

Tan, L. H., Laird, A. R., Li, K., & Fox, P. T. (2005a). Neuroanatomical correlates of phonological processing of Chinese characters and alphabetic words: a meta-analysis. *Hum Brain Mapp, 25*(1), 83-91. https://doi.org/10.1002/hbm.20134

Tan, L. H., Spinks, J. A., Eden, G. F., Perfetti, C. A., & Siok, W. T. (2005b). Reading depends on writing, in Chinese. *Proceedings of the National Academy of Sciences, 102*(24), 8781-8785.

Tian, S., & Baayen, R. H. (2022). Productivity and semantic transparency: An exploration of word formation in Mandarin Chinese. *The Mental Lexicon, 17*(3), 458-479.

Treisman, A. M. (1960). Contextual cues in selective listening. *Quarterly Journal of Experimental Psychology, 12*(4), 242-248.

Tsai, J.-L., Lee, C.-Y., Lin, Y.-C., Tzeng, O. J., & Hung, D. L. (2006). Neighborhood size effects of Chinese words in lexical decision and reading. *Language and Linguistics, 7*(3), 659-675.

Tsang, Y.-K., Zou, Y., & Tse, C.-Y. (2022). Semantic transparency in Chinese compound word processing: Evidence from mismatch negativity. *Neuroscience, 490*, 216-223.

Wang, C., & Peng, D. (1999). The roles of surface frequencies, cumulative morpheme frequencies, and semantic transparencies in the processing of compound words. *Acta Psychologica Sinica.*

Wang, Y., Li, Z., Jiang, M., Long, F., Huang, Y., & Xu, X. (2024). Time course of Chinese compound word recognition as revealed by ERP data. *Language, Cognition and Neuroscience, 39*(1), 55-75.

Wei, Y., Niu, Y., Taft, M., & Carreiras, M. (2023). Morphological decomposition in Chinese compound word recognition: Electrophysiological evidence. *Brain and Language, 241*, 105267.

Wu, J., Chang, J., Qiu, Y., & Joseph, D. (2020). The temporal process of visual word recognition of Chinese compound: Behavioral and ERP evidences based on homographic morphemes [in Chinese]. *Acta Psychologica Sinica, 52*(2), 113-127.

Xia, Q., Wang, L., & Peng, G. (2016). Nouns and verbs in Chinese are processed differently: Evidence from an ERP study on monosyllabic and disyllabic word processing. *Journal of Neurolinguistics, 40*, 66-78. https://doi.org/10.1016/j.jneuroling.2016.06.002

Xu, T. (2005). Character-based perspective and language research [in Chinese]. *Language Teaching and Linguistic Studies*(6), 5-15.

Xu, Z. (2025). Chinese Morphology. In: Oxford University Press.

Yan, G., Tian, H., Bai, X., & Rayner, K. (2006). The effect of word and character frequency on the eye movements of Chinese readers. *British Journal of Psychology, 97*(2), 259-268.

Yu, X., Tian, X., & Lau, E. (2024). Electrophysiological responses to syntactic and "morphological" structures: evidence from Mandarin Chinese. *bioRxiv, 2024*, 01-31.

Zhang, B. (1993). Decomposed storage of Chinese disyllabic words in mental lexicon [in Chinese]. *Collection of Abstracts from the 9th Chinese Psychological Conference.*

Zhang, B., & Peng, D. (1992). Decomposed storage in the Chinese lexicon. In *Advances in Psychology* (Vol. 90, pp. 131-149). Elsevier.

Zhang, D., & Koda, K. (2013). Morphological awareness and reading comprehension in a foreign language: A study of young Chinese EFL learners. *System, 41*(4), 901-913.

Zhang, K., Gu, F., & Yu, H. (2023). Early lexical processing of Chinese one-character words and Mongolian words: a comparative study using event-related potentials. *Frontiers in psychology, 13*, 1061990.

Zhang, Q., & Zhang, B. (1997). Research on concreteness effects of Chinese two-character words [in Chinese]. *Acta Psychologica Sinica, 29*, 216-224.

Zhang, Q., Zhang, J. X., & Kong, L. (2009). An ERP study on the time course of phonological and semantic activation in Chinese word recognition. *Int J Psychophysiol, 73*(3), 235-245. https://doi.org/10.1016/j.ijpsycho.2009.04.001

Zhang, W., Yan, Z., Dong, Q., Dong, J., Lu, Y., Duan, X., & Yan, H. (2024). Semantic transparency modulates the semantic perception of morphemes: Evidence from RSA of BOLD signals. *Neuroscience Letters, 823*, 137649.

Zhao, S., Wu, Y., Li, T., & Guo, Q. (2017). Morpho-semantic processing in Chinese word recognition: An ERP study. *Acta Psychologica Sinica, 49*(3). https://doi.org/10.3724/sp.J.1041.2017.00296

Zhou, X., & Marslen-Wilson, W. (1995). Morphological structure in the Chinese mental lexicon. *Language and Cognitive Processes*, *10*(6), 545-600.
Zhou, X., Ye, Z., Cheung, H., & Chen, H.-C. (2009). Processing the Chinese language: An introduction. *Language and Cognitive Processes*, *24*(7-8), 929-946. https://doi.org/10.1080/01690960903201281
Zou, L., Packard, J. L., Xia, Z., Liu, Y., & Shu, H. (2015). Neural Correlates of Morphological Processing: Evidence from Chinese. *Front Hum Neurosci*, *9*, 714. https://doi.org/10.3389/fnhum.2015.00714

Chapter 4
Methodological Approaches to Chinese Reading and Morphological Processing

In the past two decades, research on reading has undergone a radical methodological transformation. A wide range of new tools and techniques have been introduced to probe the "black box" of language perception, production, and processing. These state-of-the-art approaches, grounded in advances of cognitive sciences and neuroscience, offer increasingly sensitive measures of the temporal and spatial dynamics of brain activities during reading. With this methodological survey, this chapter aims to summarize classic behavioral paradigms, emerging neuroimaging methods, and recent statistical developments in Chinese reading research over the past twenty years. In particular, we highlight how these methods have been applied not only to Chinese word and character recognition, but also to the study of morphological processing in Chinese, which poses unique challenges and opportunities compared with alphabetic languages. Beyond the theoretical motivations outlined in earlier chapters, an equally important impetus for this book arises from observations of Chinese L2 lexical errors in learner corpora, as discussed in Sect. 4.1.3 on Child and L2 Corpora. Finally, we outline future prospects of these methodologies for advancing the broader field of reading science.

4.1 Corpus-Based Approaches

In this section, we use corpus-based approaches in a broad sense. Strictly speaking, a corpus refers to a collection of naturally occurring language materials that emphasize representativeness and authenticity of usage, whereas a normative dataset is an experimentally constructed resource providing psycholinguistic or neurolinguistic indices (e.g., ratings of valence, arousal, familiarity). For the sake of analytical coherence, this section includes both corpora and normative datasets, as both serve as systematic, large-scale resources that support the study of reading and morphological processing.

F. Gao, *The When, Where, and How of Morphological Processing in the Human Brain*, https://doi.org/10.1007/978-981-95-5981-7_4

4.1.1 Mining Patterns from Real-World Corpora

Corpus-based approaches have long been fundamental to reading research in alphabetic languages such as English. Large-scale lexical databases, including CELEX Lexical Database (Baayen et al. 1996), the British National Corpus (Consortium 2007), and the English Lexicon Project (Balota et al. 2007), have provided detailed frequency norms, neighborhood measures, and lexical decision times for tens of thousands of words. These resources have been used to identify key predictors of word recognition, such as frequency, length, and orthographic neighborhood density, and to model how multiple lexical variables jointly shape processing. For example, regression and mixed-effects modeling applied to corpus-derived measures have revealed that frequency, neighborhood size, and semantic diversity all significantly predict lexical decision and naming latencies (Balota et al. 2007; Hoffman et al. 2013). Computational approaches, including connectionist models, have further demonstrated how readers integrate orthographic, phonological, and morphological information, offering mechanistic accounts of word recognition (Plaut et al. 2020; Seidenberg and McClelland 1989). Such studies highlight the value of corpus-driven approaches in building a more comprehensive understanding of reading processes.

In Chinese reading research, a number of corpus-based databases have been developed to provide frequency as well as character- and word-level information. Examples include the Lancaster Corpus of Mandarin Chinese (McEnery and Xiao 2004), Chinese Gigaword (Parker et al. 2011), Sinica Corpus (Chen et al. 1996), SUBTLEX-CH (Cai and Brysbaert 2010), Chinese lexical database (CLD, Sun et al. 2018), and Chipola (Zhao and Lei 2025). Corpus-based analyses of Chinese uncover meaningful patterns in word structure, part of speech, frequency, lexical decision behavior, and norming data. For instance, data from the Sinica Corpus indicate that monosyllabic words represent only about 2–3% of the total word types in the lexicon, yet they account for nearly half of all word tokens in actual usage. In contrast, disyllabic words dominate in terms of type variety but occur with roughly the same overall token frequency as monosyllabic words. Additionally, analyses revealed that high-frequency words tend to exhibit part-of-speech ambiguity, with over 64% of the top 500 words being ambiguous compared to only 12.4% of lower-frequency words. Even after controlling for frequency, verbs show greater ambiguity than nouns (Huang et al. 2002). Moreover, a recent quantitative study (Hong et al. 2023) based on the Lancaster Corpus of Mandarin Chinese introduced Kolmogorov complexity as a novel measure to model structural complexity, showing that more structurally complex words tend to have lower frequency, underscoring an inverse correlation between morphological complexity and usage frequency. These findings illustrate how large-scale corpora and computational metrics can illuminate the trade-offs between structural complexity, frequency, and lexical ambiguity, providing critical constraints for designing psycholinguistic experiments.

4.1.2 Normative Databases of Psycholinguistic Variables

Unlike corpus-based databases, which emphasize the objective distribution of language use (e.g., word/character frequency, stroke number), normative databases focus on human subjective judgments or psychometric ratings of linguistic units, typically collected through behavioral experiments or questionnaires. These datasets capture participants' perceptions of variables such as familiarity, semantic transparency, concreteness, imageability, and affective dimensions (valence, arousal, dominance). In this sense, a normative dataset provides "psychological norms" that reflect how words are mentally represented and processed, offering a crucial complement to frequency-based measures from corpora.

Over the past decade, a number of large-scale psycholinguistic norms in Chinese have been developed, covering both simplified and traditional scripts (see Table 4.1). For instance, Song and Li (2021) provided ratings for 3783 two-character words on familiarity, concreteness, age of acquisition (AoA), semantic transparency, and affective features. Xu and colleagues further expanded normative resources with AoA ratings for nearly 20,000 words and concreteness ratings for 9877 two-character words (Xu and Li 2020; Xu et al. 2021). Similarly, Su et al. (2022) introduced the first large-scale imageability database for 10,426 words, showing its independent contribution to lexical decision performance beyond concreteness. These developments ensure that key psycholinguistic variables, including familiarity, AoA, concreteness, and imageability, are now broadly available for Chinese.

In addition to lexical factors, affective norms have also been widely established, specifically for Chinese emotion words. Yao et al. (2017) collected valence, arousal, concreteness, familiarity, imageability, and context availability ratings for 1100 emotional words, and Xu et al. (2022) published ratings for 11,310 simplified Chinese words, providing one of the largest affective lexicons to date. These resources are invaluable for studies of embodied cognition, emotional processing, and clinical neurolinguistics, paralleling influential affective norm databases in English (e.g., Bradley and Lang 1999; Warriner et al. 2013).

In particular, these normative datasets provide important resources and benchmarks for research on Chinese reading and morphological processing, enabling researchers to systematically manipulate or control lexical variables and compare findings across studies. However, important gaps remain. In particular, there is still a lack of large-scale normative ratings of semantic transparency in simplified Chinese, as well as systematic annotations of morphological structure (e.g., headedness, compounding types) and conceptual relations (e.g., thematic vs. attributive). The development of such datasets and toolkits would greatly advance the field by allowing finer-grained investigations of how morphological and semantic properties modulate word recognition. Emerging tools from natural language processing (NLP, Wang and Xu 2025) and large language models (LLMs) hold promising potential to

Table 4.1 Representative Normative Databases[a]

Authors (year)	Participants (n)	Materials	Psycholinguistic variables
Yao et al. (2017)	960	1100 Chinese words	Valence, arousal, concreteness, familiarity, imageability, and context availability
Xu and Li (2020)	1081	9877 two-character Chinese words	Concreteness/abstractness
Xu et al. (2021)	1765	19,716 simplified Chinese words	AoA
Song and Li (2021)	7225	3783 two-character words	Familiarity, concreteness, imageability, AoA, subjective frequency, subjective number of meanings (sNOMs), compositionality, emotional experience rating, sensory experience rating (SER), and semantic transparency
Su et al. (2022)	415	10,426 Chinese two-character words	Imageability
Xu et al. (2022)	2421	11,310 simplified Chinese words	Valence and arousal
Wu and Mu (2024)	169	1130 Chinese words	SER
Su et al. (2023)	1300	20,275 two-character, 1231 three-character, and 2819 four-character simplified Chinese words	Familiarity
Wu (2023)	211	1083 Chinese Emotion Words	Emotional Prototypicality (EmoPro)
Tsang et al. (2024b)	80	500 sentences (Traditional Chinese)	Word segmentation agreement at each character position
Li et al. (2024a)	40,000+	over 10,000 cue words	Word associations

[a]Certain megastudies also provide normative data for psycholinguistic variables, which will be introduced in Sect. 4.4.2

address these gaps (Brysbaert et al. 2024; Martinez et al. 2024; Martínez et al. 2024), as they can be leveraged to generate large-scale predictions of semantic transparency and morphological structure with high efficiency and consistency. Integrating these computational resources with human-rated norms could substantially enrich the landscape of Chinese psycholinguistic databases and foster new theoretical and empirical insights.

4.1.3 Child and L2 Corpora

Child corpora have provided foundational resources for psycholinguistic studies of Chinese and cross-linguistic comparisons. The Chinese CHILDES corpora (Tardif 1996, 1993) collected naturalistic child–adult interactions, enabling investigations of lexical development, noun–verb asymmetry, and early morphological awareness. These corpora parallel influential English resources such as Brown (1973) and Theakston et al. (2001), which have long been central to studies of child language acquisition. Such early datasets established the methodological groundwork for examining how lexical categories and morphosyntactic structures emerge in the course of language development.

Building on this foundation, recent years have seen the development of large-scale, grade-level child corpora in Chinese that directly target reading, lexical development, and literacy acquisition. Li et al. (2023) developed the Chinese Children's Lexicon of Written Words (CCLOWW), encompassing 34 million characters and 22 million words from textbooks and extracurricular materials, with grade-level frequency and contextual diversity indices. Further, Li et al. (2024b) introduced the Chinese Children's Lexicon of Oral Words (CCLOOW), which consists of 3920 unique character and 22,229 word types from 2.7 million characters and 1.8 million words drawn from animated films and TV series, capturing 3-to-9-year-old Chinese children's oral language input. These resources allow precise modeling of vocabulary growth, contextual diversity, and morphological complexity across different stages of reading development.

Beyond L1 acquisition, increasing emphasis has been placed on learners' interlanguage corpora. Resources such as the HSK Dynamic Composition Corpus (Beijing Language and Culture University), Wu et al. (2016), and Chen and Xu (2019) documented Chinese L2 learners' productions, error patterns, and morphosyntactic development across proficiency levels. These resources not only shed light on how non-native learners acquire Chinese morphology but also bridge psycholinguistic theory with applied domains such as pedagogy, curriculum design, and language assessment.

Interestingly, beyond the theoretical background discussed in Chaps. 2 and 3, another major motivation of this book comes from corpus-based observations. In particular, analyses of Chinese L2 learner corpora revealed a striking phenomenon of lexical coinage (or invention), where learners created novel word-like forms that do not exist in standard dictionaries but remain interpretable to native speakers. For example, L2 corpora showed that learners may produce 兵人 (/bing1 ren2/, literally "soldier-person") instead of the target word 军人 (/jun1 ren2/, "soldier"). Such coinages suggest that learners may not have fully established whole-word representations in their mental lexicon and instead rely on morphemic recombination. As Dong (2011) pointed out, lexical errors in Chinese interlanguage can be broadly divided into errors in word construction and errors in word usage. From the perspective of lexical representation, entries in the mental lexicon may be stored not only as words but also as morphemes, and in some cases even as idioms (Levelt 1989). Given that Chinese is a morphologically sparse language, lexical entries are often stored at either the word or morpheme level. Building on this view, we could classify learners' coinages into morpheme-level errors and whole-word level errors, corresponding to Dong's (2011) categories of errors in construction and errors in usage, respectively.

At the whole-word level, erroneous forms refer to words that do exist in the lexicon but are misused in terms of collocation, context, or register. Many of these errors stem from formal similarity, manifested as formal mis-selection or malapropisms (Fay and Cutler 1977), synforms or similar lexical forms (Laufer 1988), and confusables (Room 2013). For instance, *equivocal* instead of *equivalent* was used in "the two vectors are equivocal"). 认识 /ren4 shi2/ "to know (to know a person or to recognize)" and 知道 /zhi1 dao4/ "to know (to know a fact or piece of information)" can be mixed in Chinese context. From the perspective of lexical representation, such errors may occur because entries lack full encoding of syntactic, contextual, stylistic, or affective information, resulting in substitution or misapplication. Existing research on Chinese lexical errors has primarily focused on whole-word misuses, often attributed to the roles of orthographic distance and cross-linguistic transfer from learners' L1 (Koda et al. 1997; Odlin 2003).

Unlike whole-word errors, lexical coinages occur at the morphemic level of the mental lexicon, arising from the misuse or recombination of morphemes to form words or "word-like forms" that do not exist in the native lexicon. Dewaele (1998), drawing on Levelt's (1989) model of speech production, classified coinages in French interlanguage speech, focusing on precisely these non-whole-word phenomena. Existing studies on Chinese lexical coinages based on L2 learners' corpora have primarily provided taxonomic and quantitative analyses from a linguistic perspective (Hou 2019; Li 2013; Xing 2003), but few have described them from the viewpoint of lexical representation. We therefore reanalyzed Chinese coinages within the framework of the mental lexicon.

Specifically, we retained 234 Chinese coinages from the 520 erroneous forms (from diverse L1 backgrounds) retrieved from Chinese Interlanguage Corpus by Xing (2003), after removing noisy items like Japanese words and unclassifiable cases. These were then analyzed in terms of the internal structure of the mental lexicon (Levelt 1989), from which we derived their distributional proportions. As shown in Table 4.2, lexical coinages related to semantics and morphology at the morphemic level are the most prominent, with synonym morpheme substitutions accounting for the largest proportion of coinages (51.3%). This proportion is highly consistent with the statistics reported by Xing (2003), Li (2013), and Wang (2009). These findings also highlight the role of morphemic representation and acquisition in L2 lexical representation. According to Wang (2009), lexical coinage errors among Chinese learners arise from insufficient or incomplete lexical

Table 4.2 Statistical analysis of morphemic errors within lexical representation framework

	Coinage category	Percentage	Coinage	Target	Notes
Semantics	Synonym morpheme substitution	51.30%	兵人	军人	"兵" and "军" are synonyms ("soldier")
	Synonym blending	7.80%	花公园	花园 or 公园	Contamination between garden and park
Morphology	Morphological rule overgeneralization	38.00%	物质品	物质	Erroneous suffix usage ("品")
Phonology	Homophone morpheme substitution	2.60%	暖器	暖气	"器" and "气" are homophones (qi4)
Syntax	Same-category morpheme substitution	0.40%	反然	反之	"然" and "之" are function words

representations. For instance, synonym morpheme substitution reflects deficiencies in semantic representation, while coinages formed through morphological rules point to incomplete morphological representation. Yet, this account focuses primarily on the internal lexical knowledge encoded within entries of the mental lexicon, while overlooking its dynamic nature and the interactive relations among different lexical units.

Lexical knowledge in the mental lexicon of L2 learners is a dynamic continuum (Palmberg 1987). However, due to the limitations of available interlanguage corpora, current resources cannot precisely capture the learning stages at which learners produce whole-word errors versus morphemic errors, and therefore fail to describe the developmental trajectory of lexical representation units. At the same time, because whole-word and morpheme units interact with each other, learners' errors can be interpreted as instances where the representation of either unit "malfunctions" at a given stage of development. For example, although lexical coinages primarily occur at the morphemic level of the mental lexicon, their production and representation are closely tied to both whole-word and morphemic units. If learners have established accurate whole-word representations, morpheme-substitution coinages such as 兵人 ("soldier-person" for 军人) or 发哭 /fa1 ku1/ ("emit-cry" for 哭/ku1/, "cry") would not occur. This observation directly motivates us to investigate the morphological units of the mental lexicon in both native Chinese speakers and L2 learners from a psycholinguistic perspective.

Building on these corpus-based findings, we formulated our first research question (Chap. 5) and proceeded to explore it through psycholinguistic and neurolinguistic paradigms and techniques, which will be reviewed in the following sections.

In sum, corpus-based approaches have played a vital role in uncovering systematic patterns of Chinese word usage, structural complexity, and lexical errors, offering large-scale, ecologically valid insights grounded in native speakers' and learners' real language production. Such findings provide the empirical foundation upon which hypotheses about lexical representation and morphological processing can be built. However, while corpora reveal what native speakers and learners produce, they do not directly address the underlying cognitive mechanisms. For this reason, corpus evidence must be integrated with experimental research, where variables can be carefully manipulated and controlled to probe psycholinguistic processes such as morphological decomposition, semantic access, and lexical competition. Taken together, the combination of corpus-based observation and experimental investigation offers the most powerful pathway toward a comprehensive understanding of Chinese reading and morphological processing.

4.2 Behavioral Paradigms

4.2.1 Lexical Decision Tasks

Lexical decision paradigm is the cover term for many experiments in which participants are asked to determine whether the presented stimuli are existing lexical items, or whether they are related to other stimuli, either phonologically or semantically. This paradigm has been the most classic test in word recognition and lexical processing research since last century (Meyer and Schvaneveldt 1971). Generally speaking, it is believed to depict the process of visual analysis of word forms, phonological processing, and semantic processing, which reflect the task demands in word recognition and meaning retrieval. In word reading studies, various lexical variables were manipulated, so as to evaluate their modulations in lexical access, among which frequency and semantic transparency received the most attention (Kuo et al. 2004; Palumbo et al. 2017; Sandra 1990; Zhang 2009). Meanwhile, neuroimaging methods (e.g., EEG, fMRI) are combined with the behavioral lexical decision task, allowing researchers to record both behavioral data (reaction time and accuracy rate) and brain activity data simultaneously. These integrated approaches lend support to the neurocognitive processes of language reading.

There are different variants of classic lexical decision task, such as semantic judgment (e.g., Liu et al. 2003; Zhang et al. 2019, 2009), homophone judgment (e.g., Chen et al. 2008; Kuo et al. 2004; Zhang et al. 2009), and morphological judgment (e.g., Ip et al. 2019).

Meanwhile, the **priming paradigm** is a widely used technique in psycholinguistics, in which the processing of a target stimulus is influenced by a preceding prime. It can be combined with tasks such as lexical decision or word naming and is particularly suited for investigating morphological processing in Chinese. Different types of priming capture different aspects of linguistic representation: **form priming** arises when the prime and target share orthographic or phonological similarity (e.g., 马 "horse" → 马虎, horse-tiger, "careless"), **semantic priming** will occur when the two are meaningfully related (e.g., 医生 /yi1 sheng1/ "doctor" → 护士/hu4 shi4/ "nurse"), and **syntactic priming** refers to the facilitation of processing when a structural pattern is repeated (e.g., exposure to a double-object construction like "The boy gave the girl a book" facilitates processing of "The teacher sent the student a letter"). These effects are generally explained by automatic spreading activation in the mental lexicon (Collins and Loftus 1975) and by interactive activation within localist network models (McClelland and Rumelhart 1981).

For instance, as a subtype of form priming,[1] repetition priming uses either the whole word or one of its constitute morphemes as the prime and the whole word as the target in tasks such as lexical decision. By comparing the degree of facilitation across different lexical units, researchers can infer the representational units of the mental lexicon. For example, Peng et al. (1994) examined the recognition of Chinese compounds and monomorphemic words under repetition priming and found no significant difference between single-character and whole-word primes, supporting a decompositional account of lexical representation.

4.2.2 Word and Character Naming

Even though naming task differs from lexical decision in response modality, it has been shown that they rely on similar word recognition and lexical retrieval processes (Carreiras et al. 2007; Tsang et al. 2018). In a naming task, participants are asked to read aloud the visually presented stimuli, during which both lexical forms ("lexeme" in Levelt's account) and meanings ("lemma") are processed or retrieved to accomplish the word production. In particular, Carreiras et al. (2007) demonstrated very similar patterns of brain activation for lexical decision and naming in areas associated with word reading (e.g., left fusiform gyrus, posterior temporal cortex, pars opercularis, and bilateral insulae), but task differences were observed bilaterally in sensorimotor areas.

Character naming is a popular task for investigating the neurocognitive processes of Chinese character recognition (Lee et al. 2004; Liu and Perfetti 2003; Yum et al. 2014; Zhao et al. 2012). For example, Liu and Perfetti (2003) compared the ERP correlates of Chinese character and English word reading among Chinese bilinguals. They found that Chinese produced an earlier and higher amplitude shift (N150) than English, while frequency effects were significant for both Chinese and English at 250 ms, but at 450 ms, only English frequency effect was reliable. This line of research showed that naming task could be a viable tool for investigating the temporal unfolding of orthographic, phonological, and semantic processing, as well as the modulations of lexical variables (e.g., word frequency).

In addition, as an extension of naming paradigm, word association task also provides a time window to look at the semantic processing during word production. In the seminal work from Tan et al. (2000), a network system which included frontal (BA9/47), temporal (BA 37) cortices were found to be associated with Chinese word reading in a word association task. More recently, the Small World of Words

[1] This is the paradigm we used in Chap. 5.

Project has released a large-scale Mandarin Chinese word association database containing over 10,000 cue words, which offers rich resources for advancing psycholinguistic and neuroscientific research (B. Li et al. 2024a).

4.2.3 Eye Tracking and Mouse Tracking

Eye-tracking technology exploits the close temporal link between gaze and cognition to study the fast and highly automatic processes involved in language processing. The field of psycholinguistics has seen a surge in the application of eye-tracking techniques to the study of reading since the 1980s (Clifton et al. 2007). Given the uniqueness of Chinese script at character-layout level and morphology level (Perfetti et al. 2013), eye-tracking technique has been widely used to explore the visual analysis and meaning access when participants were reading words independently or embedded in texts. For example, Gu et al. (2015) explored how character order information is encoded in isolated word processing or Chinese sentence reading in two experiments by using a masked priming paradigm and a gaze-contingent display-change paradigm. Their results showed that character order was encoded at an early stage of processing in Chinese reading, but character position encoding was not strict. This study represented an important contribution to the understanding of character order encoding during Chinese reading.

To summarize the existing literature, eye-tracking research on Chinese word reading can be broadly grouped into two major streams. First, studies have investigated the time course of word processing by examining fixation durations, saccade patterns, and parafoveal preview effects, providing fine-grained evidence about the temporal unfolding of lexical access (e.g., Ma et al. 2019; Yan et al. 2006; Zhou et al. 2018). Second, a substantial body of work has examined how both native speakers and L2 learners segment words during naturalistic reading, highlighting the role of word boundary cues, morphological information, and contextual predictability in guiding segmentation decisions (Li et al. 2025, 2011; Ma et al. 2015; Shen et al. 2010, 2012; Zhou et al. 2018). Collectively, these research streams provide the foundation for more targeted investigations into Chinese morphological processing using eye-tracking methods.

In particular, Shen et al. (2018) provided auditory evidence that the second morpheme of disyllabic compounds is processed incrementally, suggesting that morphological effects emerge dynamically across modalities. Luo et al. (2023) further demonstrated that morphological structure influences saccade, indicating that eye movements are sensitive not only to word boundaries but also to internal morphemic organization. In parallel, Pan et al. (2023) used a gaze-contingent boundary paradigm to show that parafoveal morphological information can be accessed during reading, highlighting the role of morphological awareness in both Chinese and Korean adults. Together, these studies underscore the importance of morphological structure and awareness in shaping the temporal course and efficiency of Chinese word recognition.

More recently, mouse tracking has been introduced as a complementary method for studying online word recognition. Similar to eye movements, mouse trajectories provide a continuous index of real-time decision dynamics, but with a focus on motor responses. Lin and Lin (2020) showed that transposed-character effects in Chinese compound word recognition could be captured via trajectory deviations, revealing the graded competition between candidate words. Chen et al. (2025) extended this approach to prelingually deaf adults, demonstrating both transposed- and substituted-character effects, thereby highlighting the robustness of morphological and orthographic influences across populations. In addition, Lin et al. (2023) applied mouse tracking to spoken word recognition, disentangling the role of syllables versus phonemes in Mandarin and English, and confirming its utility beyond visual modalities. These findings illustrate that mouse tracking, in parallel with eye tracking, can reveal fine-grained dynamics of morphological and general lexical processing.

One of the key advantages of eye tracking is its high temporal sensitivity in capturing naturalistic reading behavior. Importantly, it can be integrated with neuroimaging techniques to bridge behavioral dynamics with underlying brain mechanisms (Bonhage et al. 2015; Dimigen et al. 2011; Pinheiro et al. 2024). For example, Fu et al. (2025) combined eye tracking with EEG to trace the neurocognitive processes of Spanish–English and Chinese–English bilingual novel word acquisition, mapping moment-to-moment gaze behavior onto cortical activity. Similarly, mouse tracking has also been integrated with neural measures: Chen et al. (2022) provided a resource for assessing dynamic binary choices during decision making using simultaneous EEG and mouse tracking, while Notaro and Diamond (2018) demonstrated the feasibility of combining EEG, eye tracking, behavioral, and screen-capture data (e.g., mouse movements, clicks, and keypresses) in online (i.e., the web-based Duolingo interface) German language learning. Such multimodal approaches hold promise for advancing our understanding of how morphology, orthography, and semantics are jointly processed in Chinese reading, and how these mechanisms extend to L2 and bilingual contexts.

4.2.4 Other Tasks

In addition to experimental paradigms such as lexical decision, naming, and eye tracking, a wide range of paper-and-pencil tasks and interview-based assessments have been employed to investigate morphological knowledge, particularly in children and L2 learners. These tasks are especially valuable for developmental and educational research, as they can be administered to large groups in naturalistic classroom settings.

Common paper-based tasks include morpheme judgment tasks (e.g., deciding whether a string is a real morpheme), morpheme combination tasks (e.g., combining given morphemes to form new words), and morpheme production tasks (e.g., generating words that contain a target morpheme). Such measures provide insights into learners' explicit knowledge of morphological structure and their ability to manipulate morphemes in novel contexts (Berthiaume et al. 2018; Carlisle 2000; Ku and Anderson 2003). Morphological decomposition tasks, where participants are asked to break down a word into its morphemic components, are also frequently used to probe awareness of compounding and derivation in Chinese (Shu et al. 2006).

Interview-based approaches further complement these tasks by eliciting metalinguistic reflections from learners. For instance, children or L2 learners may be asked to explain the meaning of unfamiliar compounds by inferring from constituent morphemes, or to justify why certain morpheme combinations are acceptable while others are not. These qualitative data provide additional evidence for how learners conceptualize and apply morphological rules in real-time language use (Carlisle 2000; Carlisle and Fleming 2003).

Together, these tasks expand the methodological toolkit for studying morphological knowledge processing by targeting explicit morphological awareness, offering valuable perspectives that complement reaction-time and neuroimaging measures.

4.3 Neuroimaging Methods

The rapid development and availability of neuroimaging methodologies over the past twenty years have transformed psycholinguistic research and advanced the understandings into the neural mechanisms of language comprehension. In light of our observations, the most-widely used neuroimaging modalities in reading research are fMRI and EEG, while the utilization of fNIRS and MEG in this field are still developing. Here we will briefly summarize the basic principles of these neuroimaging methods including features and strengths, as well as applications in Chinese reading studies, in particular word and character reading. The relative advantages and pitfalls concerning different methods and the prospects of technology fusion are also discussed. Some future directions of these technologies in the field of reading science are thus highlighted.

4.3.1 Electroencephalography (EEG) and Event-Related Potential (ERP)

EEG has been widely applied in reading research to capture the online neural activity underlying language comprehension. By placing electrodes on the scalp, EEG records real-time brain activity with millisecond-level temporal resolution, allowing researchers to determine precisely when cognitive operations occur during reading. To extract meaningful signals from the continuous neural activity, ERPs are derived, providing time-locked responses to specific stimuli that can be linked to functions such as attention, emotion, and language.

After preprocessing the raw EEG data, which typically includes re-referencing, band-pass filtering, epoch segmentation, and artifact rejection, ERP waveforms can be extracted. Averaging across multiple trials further improves the signal-to-noise ratio, yielding reliable indicators of cognitive processes. ERP waveforms manifest as positive or negative voltage deflections, known as ERP components, which are conventionally characterized by polarity (e.g., P300 vs. N400), peak latency (e.g., N400 peaking around 400 ms after word onset), and scalp distribution (e.g., Left Anterior Negativity, LAN). The procedure of EEG acquisition and preprocessing is illustrated in Fig. 4.1. During reading or listening tasks, participants' scalp electrical activity is recorded and amplified by the EEG system, while linguistic materials are time-locked with event markers that correspond to specific stimulus types (e.g., different lexical categories). These markers are synchronized with the EEG signal, enabling the extraction of event-related responses. The continuous EEG is then preprocessed offline and averaged by event type, resulting in ERPs. This method is particularly well-suited for revealing the fine-grained temporal dynamics of language processing.

The most common ERP components of interest to reading research are N200, P200, N400, and P600. For instance, N400 is a negative-going ERP component that typically peaks around 400 ms after stimulus onset. It is most prominent over centro-parietal scalp regions and is sensitive to semantic integration difficulty. Functionally, the N400 reflects processes of meaning access and integration, being larger when a word or stimulus is unexpected or incongruent in context. By contrast, the P600 is a positive-going component that peaks around 600 ms and is typically maximal over posterior scalp regions. It is most strongly elicited by syntactic anomalies, garden-path sentences, or reanalysis processes, and has therefore been interpreted as reflecting syntactic repair, structural reanalysis, or more generally the

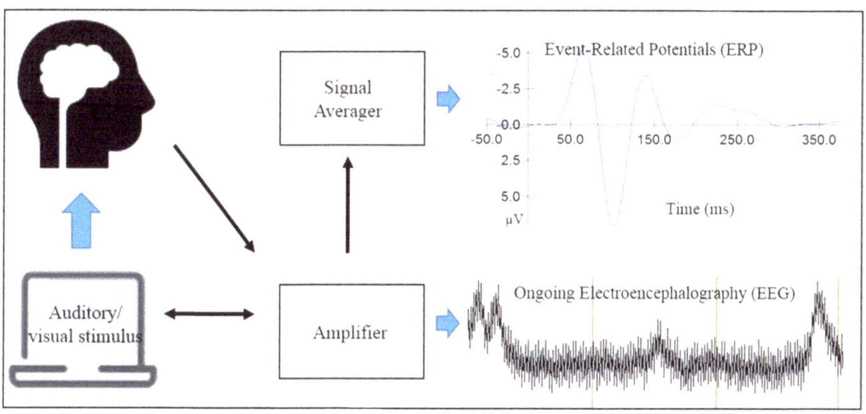

Fig. 4.1 The schematic demonstration of EEG acquisition and ERP preprocessing

reallocation of cognitive resources to resolve conflict in sentence comprehension (see more details and more ERP components in Chap. 3 of Luck 2014).

In light of existing studies, Chinese word reading implicates a general temporal unfolding of early orthographic and/or phonological processing (N170, P200; e.g., Hsu et al. 2009; Wu et al. 2020; Zhang et al. 2009, 2020) and then later semantic processing (N400, e.g., Du et al. 2013; Liu et al. 2003; Yum et al. 2014; Zhang et al. 2020), which is similar to alphabetical reading (Barnea and Breznitz 1998; Hauk et al. 2006). Interestingly, there might also exist an orthographic signature in the temporal unfolding of Chinese word reading (i.e., centro-parietal N200; e.g., Du et al. 2013; Zhang et al. 2012a; Zhou et al. 2016), which was thought to reflect the specific visual demands of Chinese script. However, it merits further examinations.

Beyond scalp-recorded EEG, intracranial electroencephalography (iEEG) provides a powerful window into the neural basis of language due to its exceptional spatio-temporal resolution. Because of its invasive nature, however, iEEG is typically conducted only in clinical populations. Recent work has demonstrated its value in disentangling core aspects of language processing. Employing high-density electrocorticography (ECoG) in the left IFG of ten patients undergoing awake language mapping during brain tumor surgeries, Zhu et al. (2022) revealed distinct spatiotemporal patterns of neural activity associated with syntactic and semantic processing in Chinese sentences, suggesting that syntactic operations may precede semantic interpretation. Such evidence illustrates how iEEG complements scalp-based ERP findings by offering direct cortical recordings, thereby refining our understanding of the temporal architecture of Chinese word and sentence processing. Taken together, the accumulated EEG and iEEG findings underscore the crucial role of electrophysiological methods in mapping the time course of orthographic, phonological, semantic, and syntactic processes in reading.

4.3.2 Magnetoencephalography (MEG)

Like EEG, which records scalp electrical potentials arising from synchronized postsynaptic activity, MEG is a functional brain mapping technique that non-invasively captures neurophysiological activity. However, instead of electrical potentials, MEG directly measures the magnetic fields at the scalp generated by the synchronized electrical activity of neuronal populations. Among the most actively sought applications of MEG has been localization of language processing cortex (Frye et al. 2009). With its combination of high temporal resolution and relatively precise spatial resolution, MEG is particularly well suited to examining the dynamic brain processes underlying reading and speech production. Event-related designs using single words and lexical bundles have been especially informative, helping to characterize the fundamental building blocks of language and to delineate the successive stages of processing, from orthographic analysis to semantic and syntactic integration (Pylkkänen et al. 2006; Valaki et al. 2004; Westerlund et al. 2015).

Nevertheless, as MEG remains less widely used and studied (due to the high cost) as compared to other imaging tools, MEG investigations in Chinese reading has just initiated in the past decade (Hsu et al. 2011, 2014, 2019; Hung et al. 2014; Xu et al. 2019). For example, Hsu et al. (2019) evaluated effects of word-internal variables on reading Chinese disyllabic words by manipulating morphological complexity (the number of morphemes) and structural complexity (whether the two syllables relate to each other via the specifier-head-complement structure). These results suggested that morphological complexity and structural complexity substantially modulate activities in left temporal cortex during the reading of disyllabic words. In general, MEG could imply both temporal and spatial information of word reading simultaneously. Hopefully, this technique could be more widely used to investigate the neurocognitive processes in future Chinese reading studies.

4.3.3 Functional Near-Infrared Spectroscopy (fNIRS)

FNIRS has become an increasingly favored tool for investigating brain function, owing to its non-invasive nature, lack of radiation, relatively lower cost, and portability. Using optical techniques, it detects changes in oxygenated and deoxygenated hemoglobin concentrations, as well as oxygen saturation, that accompany neural activity. A key advantage of fNIRS is its suitability for measuring hemodynamic responses in various settings with high ecological validity. For example, recordings can be conducted in naturalistic environments without physical restraint or sedation. To date, fNIRS has been applied across a wide range of domains in both typical and clinical populations, including hyper-scanning to study social interaction, neurofeedback, brain–computer interface (BCI) applications, and investigations of brain activity during both task performance and resting states (e.g., Lu et al. 2010; Wang et al. 2018; Xu et al. 2020).

Although fNIRS has been widely applied in domains such as social cognition, attention, and BCI, as well as in research on other languages (Kovelman et al. 2009; Soltanlou et al. 2018; Sugiura et al. 2011), its use in studies of Chinese reading remains relatively limited (Chen et al. 2008; Hu et al. 2018; Zhang et al. 2016). For example, Chen et al. (2008) employed a homophone judgment task to compare hemodynamic responses to Chinese characters and English words, and found differential activations in left MFG, left STG, and left supramarginal gyrus. These results converge with earlier fMRI findings (Tan et al. 2001, 2000), suggesting that distinct neural mappings are shaped by whether written language is learned in an alphabetic or logographic system. Given its high ecological validity, fNIRS holds considerable promise for investigating naturalistic reading, where traditional imaging techniques face practical constraints. Future research employing fNIRS in this domain may provide deeper insights into the neural underpinnings of reading, particularly in logographic languages such as Chinese.

4.3.4 Functional Magnetic Resonance Imaging (fMRI)

The fMRI technique relies on the principle that cerebral blood flow is tightly coupled with neuronal activity, allowing brain activation to be inferred from blood oxygenation changes. In general, increased activity in a brain region is accompanied by greater blood supply to that area. A similar neurovascular coupling mechanism underlies fNIRS, which likewise infers neural activity from hemodynamic responses, albeit with different measurement techniques. The earliest fMRI studies on language processing were reported by McCarthy et al. (1993), Hinke et al. (1993) and Price (2012). In particular, McCarthy et al. (1993) demonstrated that word generation, relative to repetition, activated the inferior frontal cortices and anterior insula. With three decades' developments, fMRI has become the tool of choice for the language and brain community. This technique has grown largely because of its high spatial resolution, non-invasiveness, and relative ease of implementation. The fMRI signal is robust and for the most part, highly reproducible and reliable. When compared to other neuroimaging methods, fMRI overcomes the poor spatial resolution of EEG, the high production-cost of radioisotopes for Positron Emission Computed Tomography (PET) studies as well as the ethical limitations for their use.

Importantly, departing from alphabetic systems, Chinese characters, rather than letters, are the building blocks of Chinese words. In most cases, each Chinese character corresponds to both a syllable and a morpheme. To map this linguistic idiosyncrasy to brain network systems, fMRI has been widely used in the past 20 years. Tan and his lab members (Tan et al. 2000, 2001, 2005b) have worked on a set of fMRI studies, which suggested that left MFG (BA9) might be the crucial area for Chinese reading. This network was then tested by following studies (Li et al. 2006; Zhou et al. 2020). Meanwhile, another line of fMRI studies has discussed the brain representations at the word level (e.g., Feng et al. 2020; Ip et al. 2019; Yang et al. 2017; Zhang et al. 2012b, 2019).

In addition, an important trend in recent years is the growing use of functional connectivity analyses as a means to investigate the functional organization of the human brain. Unlike traditional activation-based approaches, connectivity methods allow researchers to capture the interactions among distributed brain regions, offering a network-level perspective on cognitive processing. This approach has been increasingly applied in both fNIRS (e.g., Hu et al. 2018; Zhang et al. 2010) and fMRI studies (Su et al. 2025; Zhang et al. 2019), and is particularly relevant for reading research, where multiple components (e.g., orthographic, phonological, semantic, and executive systems) must be dynamically integrated across widespread cortical networks.

4.3.5 Methodological Comparisons and Multimodal Integration

We have surveyed some of the most common neuroimaging techniques employed in language comprehension and Chinese reading studies. In addition to these widely used methods, there exist other techniques that are less frequently applied in basic research but have considerable value in clinical practice, such as PET (Klein et al. 2001). Moreover, neurostimulation methods, including transcranial magnetic stimulation (TMS) (e.g., Zhang et al. 2019), transcranial direct current stimulation (tDCS) (e.g., Tao et al. 2021), focused ultrasound (FUS) (Chen et al. 2021), and more recently transcranial photobiomodulation (tPBM) (M. Yang et al. 2024a), have provided powerful tools for probing causal mechanisms in language processing and learning when used alone or in combination with neuroimaging. Another promising direction is the use of neurofeedback to regulate Chinese language learning in special populations and L2 learners. While some studies have already demonstrated its potential in English as a second language learning (e.g., Chang et al. 2021), similar applications in Chinese remain largely unexplored.

Generally, each neuroimaging method presents its own strengths and limitations, which are typically assessed along four dimensions: temporal resolution, spatial resolution, invasiveness, and financial cost. From the perspective of laboratory establishment and management, the choice of which technique to purchase and maintain is often a practical balance across these dimensions. From a research perspective, however, the most critical factor is how a given technique performs in terms of spatio-temporal characteristics (Sejnowski et al. 2014). For instance, EEG and MEG provide millisecond-level temporal resolution, making them well suited to capturing the fast dynamics of Chinese word recognition, such as the N400 component reflecting semantic integration. In contrast, fMRI offers millimeter-level spatial resolution, enabling precise localization of cortical networks underlying orthographic, semantic, and syntactic processing in Chinese reading. Nonetheless, each technique has inherent limitations: EEG suffers from relatively poor spatial resolution due to the inverse problem of source localization; fNIRS is restricted to cortical surface regions and has lower spatial resolution compared to fMRI; and fMRI, while offering high spatial precision, is limited by its low temporal resolution and sensitivity to motion artifacts. Thus, for studies of reading, the first step in methodological design is to determine which aspects of temporal dynamics or spatial localization are most relevant to the research question, alongside considerations of equipment availability.

To overcome the limitations of individual techniques and maximize their complementary strengths, multimodal integration has become increasingly prevalent in the past decade. For instance, concurrent recording with fMRI and EEG provides both high spatial and high temporal resolution, offering a richer picture of the reading process (e.g., Dogil et al. 2004; Steinbeis and Koelsch 2008). In Chinese reading studies, such methodological integration is still at an exploratory stage. Zhang et al. (2019), for example, combined fMRI with TMS to investigate the neural basis of semantic cognition in Mandarin, demonstrating how multimodal approaches can bridge correlational and causal evidence. In line with this trend, Chaps. 6–8 of this book employ a combined fNIRS–EEG approach to capture both the temporal and spatial characteristics of Chinese morphological processing.

Another emerging trend involves the development of large-scale open-access multimodal databases that combine linguistic, eye-tracking, and neuroimaging data. These databases enable cross-linguistic comparison, computational modeling, and reproducible science. For example, the Multilingual Eye-movement Corpus (MECO) has facilitated large-scale cross-linguistic investigations of reading (Siegelman et al. 2022). Similar efforts are underway in Chinese: Zhang et al. (2022) released a comprehensive dataset of eye-movement measures of words during Chinese sentence reading, while Mou et al. (2024) developed ChineseEEG, an open EEG and eye-tracking dataset for Chinese semantic alignment and neural decoding. In addition, an fMRI dataset (Wang et al. 2022) has been shared in which 11 participants thought of 672 individual concepts, spanning both concrete and abstract categories, each annotated with 54 semantic features (sensory, motor, spatial, temporal, affective, social, cognitive). This resource provides high-quality cognitive and neuroimaging data ideally suited to studying how the brain represents semantic features and conceptual knowledge, and thus contributes to bridging behavioral, linguistic, and neural levels of evidence. More recently, Tan and Zhang (2025) developed a publicly available dataset comprising behavioral and EEG recordings from 87 native Mandarin speakers across four spoken word production experiments. This resource captures the temporal dynamics of syllable and phoneme processing, providing robust evidence for the syllable effect and offering a valuable basis for cross-linguistic comparisons in phonological encoding during word production. These resources highlight the potential of open multimodal data to accelerate our understanding of the mechanisms of Chinese reading.

Beyond multimodal neurocognitive tools and datasets, genetic research provides a biological perspective on the foundations of language. The most well-known case is FOXP2, discovered in the KE family with speech and language deficits (Lai et al. 2001), which regulates neuronal migration and synaptic plasticity. Another key candidate is CNTNAP2, which has been linked to language impairment and dyslexia (Vernes et al. 2008), and together with FOXP2 forms a pathway that provides a mechanistic link across clinically distinct syndromes involving disrupted language. Recent work emphasizes a multi-gene, multi-pathway model, where language ability arises from numerous genes affecting brain development, connectivity, and plasticity, as well as strong interactions between gene and environment, such as the role of family input. Neuroimaging evidence further connects these genes to classical language regions (e.g., Broca's and Wernicke's areas, parietal cortex, basal ganglia). Importantly, Wong et al. (2022) showed that in Chinese bilinguals learning a third language (L3), genetic effects (e.g., FOXP2, CNTNAP2, DCDC2) were robust for L1 but weaker for foreign languages (especially L3), where non-genetic factors such as motivation and musical training predominated. Collectively, these findings indicate that there is no single "language gene"; instead, language and reading abilities arise from the interplay of multiple genes. The processing demands of language learning vary not only across developmental stages but also across linguistic features and language types, with different neurogenetic factors supporting different aspects of language. Moreover, these genetic influences interact with environmental input, jointly shaping the neurocognitive basis of language. Integrating genetic findings with behavioral, electrophysiological, and neuroimaging evidence is crucial for building a multi-level account of reading, particularly for elucidating the unique demands of Chinese morphological processing and its contribution to both universal and language-specific brain mechanisms.

Moving forward, future research would greatly benefit from the development of multimodal, large-scale databases that capture Chinese-specific linguistic features across multiple levels of processing (from characters to discourse). Extending such resources to include diverse populations, such as children, L2 learners, and clinical groups, will not only enrich theoretical models but also inform educational practices and clinical interventions. Ultimately, such cross-level integration will advance reading science by linking genetic, neural, cognitive, and cultural perspectives into a unified framework.

4.4 Experimental Design and Statistical Analysis

The statistical approaches used in psycholinguistics have evolved alongside the field itself. While factorial designs and ANOVA have long provided a foundation, the recognition that lexical variables are often continuous and interdependent has led to broader adoption of regression and mixed-effects models. The availability of large-scale datasets has further enabled megastudy designs and more sophisticated multivariate or machine learning approaches, while meta-analyses quantitively synthesize findings across studies. These methods reflect a shift from traditional experimental designs toward data-intensive and computationally powerful approaches in reading research.

4.4.1 Factorial Designs and ANOVA

Due to the relative ease of variable manipulation and straightforward data interpretation, factorial designs coupled with analyses of variance (ANOVA) have long been the backbone of psycholinguistic experimental research. This approach has been particularly valuable in the formative years of the field, as it allows researchers to manipulate one or more independent variables and test their main and interaction effects in a transparent way. Indeed, a large-scale methodological synthesis reported that 56% of 606 quantitative studies in *Language Learning* and *Studies in Second Language Acquisition* employed one or more ANOVA procedures (Plonsky 2013), attesting to the centrality of this method in shaping research practices.

At the same time, factorial designs are not without limitations. One issue lies in the reliance on relatively small and tightly controlled stimulus sets, which can constrain the generalizability of findings. For example, Myers (2006) argued that inconsistent results in Chinese compound word processing studies may be partly due to such methodological constraints. Tse et al. (2017) further pointed out two inherent challenges: first, many lexical variables are intercorrelated, making it difficult to manipulate one factor in isolation (e.g., frequency often co-varies with semantic transparency or visual complexity); second, lexical variables are typically continuous rather than categorical, and dichotomizing them (e.g., high vs. low frequency) may reduce statistical power (Maxwell and Delaney 1993) and obscure graded effects across the full distribution (Baayen 2004).

In the context of Chinese reading research, these challenges are particularly salient, as the logographic script and complex morphological structure make lexical variables highly interdependent. Nevertheless, factorial designs remain a useful and widely adopted tool, especially when experimental control and interpretability are prioritized. Recognizing their limitations, however, has motivated researchers to explore complementary methods that can better accommodate the multidimensional nature of lexical variables, which is an important trend in the methodological development of reading science.

4.4.2 Regression, Mixed-Effects Models, and Megastudies

Nearly half a century ago, Cohen (1968) argued that multiple regression, as the parent procedure of ANOVA, could provide not only the same information as ANOVA but also a number of improvements. Plonsky and Oswald (2017) further reinforced this view, emphasizing that regression can do everything ANOVA can do, and more. In particular, regression allows the incorporation of continuous predictors and yields interpretable results across the full range of values rather than dichotomized categories. Despite these evident advantages, regression has historically been underused in psycholinguistic and reading research, where factorial designs have remained dominant. Nonetheless, recent work has started to apply regression-based approaches more broadly. Building upon regression, mixed-effects models have become increasingly influential in both first and second language reading research. These models incorporate random effects of participants and items, thereby addressing the issue of non-independence in psycholinguistic data (Baayen et al. 2008). Particularly, mixed-effects modeling has been widely adopted in second language acquisition research to account for learner variability across tasks (see Linck and Cunnings 2015 for a review). In Chinese reading studies, mixed-effects models have been applied to eye-tracking, ERP, and behavioral datasets, providing more generalizable results by simultaneously considering fixed effects of lexical variables (e.g., frequency, transparency) and random variation at the subject and item levels (e.g., Law et al. 2017; Lee et al. 2015; Yi 2018). By capturing this hierarchical structure, mixed-effects approaches offer a more powerful and flexible framework than traditional ANOVA.

In parallel, the rise of megastudy approaches represents another important methodological development. Corpus-driven megastudies combine lexical databases with large-scale behavioral data (e.g., lexical decision times, naming latencies) to identify predictors of processing efficiency across thousands of items (Keuleers and Balota 2015). As Balota et al. (2012a, 2012b) noted, megastudies allow the language to define the stimuli, as opposed to researchers preselecting small sets of items matched on limited criteria. By leveraging agreed-upon lexical databases, researchers can examine the influences and interrelationships among multiple variables in a continuous rather than categorical manner, thereby offering more ecologically valid tests of theoretical models.

The last two decades have seen a growing body of megastudies in Chinese reading research across different levels (character-level and word-level), tasks (lexical decision and naming), and scripts (simplified and traditional Chinese) (Chang et al. 2016; Lee et al. 2015; Liu et al. 2007; Sze et al. 2014; Tsang et al. 2018; Tse et al. 2017; see Table 4.3). These studies have not only advanced conventional behavioral research but also verified the psycholinguistic reality of key lexical variables, thereby deepening our understanding of the cognitive mechanisms underlying Chinese reading (e.g., Sze et al. 2015; Y.-K. Tsang et al. 2024a; Tse et al. 2024).

Most recently, megastudy designs have extended into the electrophysiological domain. Tsang and Zou (2022) reported the first ERP megastudy of Chinese word recognition, which collected ERP data for thousands of two-character words. Their findings demonstrated systematic relationships between lexical variables (e.g., frequency, semantic transparency, concreteness) and ERP components, thereby bridging behavioral and neural evidence on a large scale. This pioneering work illustrates the potential of megastudy designs to transform neurolinguistic research, particularly in languages such as Chinese where the interaction of orthographic, morphological, and semantic factors is especially complex.

4.4.3 Multivariate Approaches

While regression and mixed-effects models have advanced beyond traditional ANOVA by allowing continuous predictors and handling crossed random effects, they typically examine the effect of each predictor in relative isolation. Multivariate approaches, in contrast, analyze multiple correlated predictors simultaneously, offering a more holistic view of the interplay among various variables. This is particularly useful in reading research, where many lexical factors (e.g., frequency, transparency, orthographic complexity) are inherently interdependent. For instance, Liu and Xiao (2024) applied supervised component regression and random forests to L2 Chinese compound processing and demonstrated that orthographic, phonological, and semantic cues jointly contribute to recognition, with L2 learners relying more on orthographic strategies compared to native speakers.

Table 4.3 Megastudies in Chinese reading research

Author (year)	Task	Participants	Materials	Lexical variables
Liu et al. (2007)	Word naming	Native mainland Chinese speakers (n = 480)	2423 Chinese single-character words	Word frequency, cumulative frequency, homophone density, phonological frequency, age of learning, age of acquisition, number of word formations, number of meanings, number of components, number of strokes, familiarity, concreteness, imageability, regularity, and initial phoneme
Sze et al. (2014)	Lexical decision	Native mainland Chinese speakers (n = 35)	2500 single characters	Frequency
Lee et al. (2015)	Lexical decision	Native Chinese speakers from Taiwan (n = 180)	3423 Chinese phonograms	Frequency, feedforward consistency, feedback consistency, homophone density, familiarity, number of strokes, radical combinability (phonetic and semantic), residual semantic ambiguity, morphological boundedness, noun-to-verb ratio
Chang et al. (2016)	Word naming	Native Chinese speakers (n = 140)	3314 traditional Chinese phonograms	Frequency, regularity, familiarity, consistency, number of strokes, homophone density, semantic ambiguity rating, phonetic combinability, semantic combinability, and the number of disyllabic compound words formed by a character
Tse et al. (2017)	Lexical decision	Native Cantonese speakers in Hong Kong (n = 594)	25,000+ traditional Chinese two-character compound words	Word and character frequency, semantic transparency, number of strokes
Tsang et al. (2018)	Lexical decision	Native mainland Chinese speakers (n = 504)	1020 one-character, 10,022 two-character, 949 three-character, and 587 four-character simplified Chinese words	Word length, number of strokes, frequency, contextual diversity, character neighborhood, number of meanings, number of pronunciations
Tsang and Zou (2022)	Go/no-go lexical decision task	Native Cantonese speakers (n = 51)	1020 two-character words	Number of strokes, frequency, contextual diversity, character neighborhood, number of pronunciations, number of homophones, concreteness, semantic transparency

Beyond behavioral studies, multivariate techniques are increasingly applied in neuroimaging research. Methods such as multivariate pattern analysis (MVPA) and representational similarity analysis (RSA) allow researchers to capture distributed neural representations of linguistic units and test how orthographic, phonological, and semantic dimensions are encoded in the brain (Dong et al. 2021; Haxby et al. 2014; Y. Yang et al. 2024b). Thus, multivariate approaches complement regression-based methods by uncovering not only the independent effects but also the interactive and hierarchical contributions of lexical properties across behavioral and neural domains. These developments suggest a promising direction for integrating psycholinguistic modeling with cognitive neuroscience of reading.

4.4.4 Meta-Analytic Approaches

While regression, mixed-effects, and multivariate methods focus on modeling individual datasets, meta-analytic approaches address a different but equally crucial question: what can be concluded across multiple existing studies? When several independent studies investigate the same research question, meta-analysis provides a powerful statistical framework for synthesizing their findings. A key benefit of this approach lies in its ability to aggregate evidence, thereby increasing statistical power and yielding more robust and generalizable estimates than those derived from any single study.

In the neuroimaging field, meta-analytic methods such as activation likelihood estimation (ALE) have been increasingly employed to determine convergence across brain regions reported by different functional imaging studies (Eickhoff et al. 2012; Murphy et al. 2019). Several meta-analyses have been conducted to identify the neural correlates of Chinese word reading (Tan et al. 2005a; Wu et al. 2012; Zhao et al. 2017; Zhou and Shu 2017). For example, Tan et al. (2005a) synthesized brain activation data from six mapping studies of Chinese word reading and concluded that the left dorsolateral frontal area (MFG, BA9) may be specific to the addressed phonology in Chinese compared to alphabetic languages. However, more recent meta-analytic work (Zhao et al. 2017) challenged this view, suggesting that previously identified script-specific regions, such as the left MFG for Chinese (Tan et al. 2005a) and the pSTG for alphabetic reading (Paulesu et al. 2000), may reflect task dependence rather than script selectivity. These debates highlight the continuing need for meta-analytic evidence in resolving long-standing questions about script specificity in reading.

Despite these advances, it is noteworthy that no meta-analyses have yet been conducted on morphological processing, either from the perspective of neural mechanisms or specifically within the domain of Chinese. Given that morphology plays an important role in reading across languages, synthesizing evidence through meta-analytic methods could provide critical insights into its neural underpinnings and clarify whether observed patterns are universal or script-specific. Future research could thus expand the scope of meta-analysis to systematically evaluate morphological processing, helping to establish a more comprehensive picture of its cognitive and neural basis, as well as its cross-linguistic variability.

4.5 Summary and Future Directions

This methodological review has surveyed the corpora, behavioral paradigms, neuroimaging techniques, and statistical tools employed in Chinese reading research over the past two decades, with particular attention to studies addressing Chinese morphological processing where available. Behavioral paradigms such as lexical decision and naming have been the most widely used, while EEG and fMRI dominate the neuroimaging landscape, with fNIRS, MEG, neuro-intervention, and genetics gradually gaining attention. At the same time, methodological advances in experimental design and statistical modeling (e.g., regression, mixed-effects models, megastudies, multivariate analyses, and meta-analyses) have substantially enriched the field. Together, these developments have paved the way for a more comprehensive understanding of Chinese reading processes.

Looking forward, we highlight three directions for future research. First, integration of behavioral and neuroimaging methods will be critical. Combining behavioral paradigms with multimodal brain imaging (e.g., EEG–fMRI, EEG–fNIRS) can provide complementary temporal and spatial resolution, offering a more fine-grained account of the mechanisms underlying Chinese reading. This integration is particularly relevant for Chinese morphological processing, where the interplay of orthographic, phonological, and semantic factors requires converging evidence from multiple methods.

Second, longitudinal and developmental designs are urgently needed to track how reading and morphological processing evolve across the lifespan. Most current studies focus on adult readers, but understanding how children and L2 learners acquire Chinese morphological knowledge requires long-term data. Developmental trajectories can shed light on how morphological awareness and processing contribute to literacy acquisition, as well as how proficiency, cognitive factors (e.g., working memory), and neural plasticity shape reading processes.

Third, open science, data sharing, and cross-linguistic comparison will play a central role in the next stage of research. Building and sharing standardized normative databases, including frequency norms, semantic transparency ratings, and morphological structure annotations, will be indispensable for reproducibility and comparability across studies. While English and other alphabetic languages already have well-established resources, Chinese still lacks large-scale, open-access datasets targeting morphological structure and semantic transparency. Moreover, cross-linguistic and cross-structural comparisons can clarify whether observed effects in Chinese reading are language-specific or universal. Finally, greater data transparency and multi-site collaboration will help overcome issues of replicability and support cumulative progress in the science of reading (Jin et al. 2023; Mackey and Porte 2012; Marsden et al. 2018).

In sum, future research on Chinese reading, and especially on Chinese morphological processing, should move toward methodological integration, developmental perspectives, and open science practices. These directions will not only deepen our understanding of the cognitive and neural mechanisms of reading but also provide practical implications for education, second language learning, and clinical applications. In the following chapters, we build on this methodological foundation by presenting a series of empirical studies that integrate behavioral paradigms with multimodal neuroimaging (EEG–fNIRS).

Specifically, Chap. 5 investigates the representation of morphological units in the Chinese mental lexicon, comparing evidence from both L1 and L2 learners. Chapter 6 examines how the brain encodes morphological constraints in Chinese compound word reading, highlighting the neural dynamics that guide lexical access. Chapter 7 focuses on the neural basis of morphological structure effects (compound vs. derivation), exploring how structural configurations modulate reading at both cognitive and neural levels. Finally, Chap. 8 extends the inquiry to bilingual contexts, uncovering the neural correlates of morphological processing in the Chinese–English bilingual brain. Together, these chapters aim to provide converging evidence for the cognitive and neural bases of Chinese morphological processing, thereby addressing key methodological and theoretical questions outlined in this review.

Through these four empirical studies, the book seeks to advance our understanding of Chinese morphological processing across theoretical, methodological, and applied dimensions, offering new insights into the nature of the mental lexicon and the neural architecture of reading. By integrating evidence from L1, L2, and bilingual contexts, it further contributes to cross-linguistic comparisons, highlighting both universal mechanisms and language-specific features of morphological processing.

References

Baayen, R. H. (2004). Statistics in psycholinguistics: A critique of some current gold standards. *Mental lexicon working papers*, *1*(1), 1-47.

Baayen, R. H., Davidson, D. J., & Bates, D. M. (2008). Mixed-effects modeling with crossed random effects for subjects and items. *Journal of Memory and Language*, *59*(4), 390-412.

Baayen, R. H., Piepenbrock, R., & Gulikers, L. (1996). The CELEX lexical database (cd-rom). Philadelphia: Linguistic Data Consortium, University of Pennsylvania.

Balota, D. A., Yap, M. J., Hutchison, K. A., & Cortese, M. J. (2012a). Megastudies: Large scale analysis of lexical processes. *Visual word recognition*, *1*, 90-115.

Balota, D. A., Yap, M. J., Hutchison, K. A., & Cortese, M. J. (2012b). Megastudies: What do millions (or so) of trials tell us about lexical processing? In *Visual Word Recognition Volume 1* (pp. 90-115). Psychology Press.

Balota, D. A., Yap, M. J., Hutchison, K. A., Cortese, M. J., Kessler, B., Loftis, B., Neely, J. H., Nelson, D. L., Simpson, G. B., & Treiman, R. (2007). The English lexicon project. *Behavior research methods*, *39*(3), 445-459.

Barnea, A., & Breznitz, Z. (1998). Phonological and orthographic processing of Hebrew words: electrophysiological aspects. *The Journal of Genetic Psychology*, *159*(4), 492-504.

Berthiaume, R., Bourcier, A., & Daigle, D. (2018). Morphological processing tasks and measurement issues. In *Morphological processing and literacy development* (pp. 48-87). Routledge.

Bonhage, C. E., Mueller, J. L., Friederici, A. D., & Fiebach, C. J. (2015). Combined eye tracking and fMRI reveals neural basis of linguistic predictions during sentence comprehension. *Cortex, 68*, 33-47.

Bradley, M. M., & Lang, P. J. (1999). *Affective norms for English words (ANEW): Instruction manual and affective ratings* (Vol. 30, No. 1, pp. 25-36). Technical report C-1, the center for research in psychophysiology, University of Florida.

Brown, R. (1973). *A first language: The early stages*. Harvard University Press.

Brysbaert, M., Martínez, G., & Reviriego, P. (2024). Moving beyond word frequency based on tally counting: AI-generated familiarity estimates of words and phrases are a better index of language knowledge. *Preprint*. https://doi.org/10.31234/osf.io/kgevy

Cai, Q., & Brysbaert, M. (2010). SUBTLEX-CH: Chinese word and character frequencies based on film subtitles. *PLoS One*, *5*(6), e10729.

Carlisle, J. F. (2000). Awareness of the structure and meaning of morphologically complex words: Impact on reading. *Reading and Writing*, *12*(3), 169-190.

Carlisle, J. F., & Fleming, J. (2003). Lexical processing of morphologically complex words in the elementary years. *Scientific Studies of Reading*, *7*(3), 239-253.

Carreiras, M., Vergara, M., & Perea, M. (2007). ERP correlates of transposed-letter similarity effects: are consonants processed differently from vowels? *Neurosci Lett*, *419*(3), 219-224. https://doi.org/10.1016/j.neulet.2007.04.053

Chang, M., Ando, H., Maeda, T., & Naruse, Y. (2021). Behavioral effect of mismatch negativity neurofeedback on foreign language learning. *PLoS One*, *16*(7), e0254771.

Chang, Y. N., Hsu, C. H., Tsai, J. L., Chen, C. L., & Lee, C. Y. (2016). A psycholinguistic database for traditional Chinese character naming. *Behav Res Methods*, *48*(1), 112-122. https://doi.org/10.3758/s13428-014-0559-7

Chen, H., & Xu, H. (2019). Quantitative linguistics approach to interlanguage development: A study based on the Guangwai-Lancaster Chinese Learner Corpus. *Lingua*, *230*, 102736.

Chen, H. C., Vaid, J., Bortfeld, H., & Boas, D. A. (2008). Optical imaging of phonological processing in two distinct orthographies. *Exp Brain Res*, *184*(3), 427-433. https://doi.org/10.1007/s00221-007-1200-0

Chen, K.-J., Huang, C.-R., Chang, L.-P., & Hsu, H.-L. (1996). Sinica corpus: Design methodology for balanced corpora. *Language, Information and Computation: Selected Papers from the 11th Pacific Asia Conference on Language, Information and Computation: 20-22 December 1996, Seoul.*

Chen, K., Wang, R., Huang, J., Gao, F., Yuan, Z., Qi, Y., & Wu, H. (2022). A resource for assessing dynamic binary choices in the adult brain using EEG and mouse-tracking. *Scientific Data*, *9*(1), 416.

Chen, Y., Huang, J., & Cheng, K. (2025). Transposed-and substituted-character effects in written word recognition by Chinese prelingually deaf adults: Evidence from mouse-tracking technology. *Journal of Experimental Psychology: Learning, Memory, and Cognition.*

Chen, Y., Li, Y., Du, M., Yu, J., Gao, F., Yuan, Z., & Chen, Z. (2021). Ultrasound neuromodulation: integrating medicine and engineering for neurological disease treatment. *BIO Integration*, *2*(4), 169.

Clifton Jr, C., Staub, A., & Rayner, K. (2007). Eye movements in reading words and sentences. *Eye movements*, 341-371.

Cohen, J. (1968). Multiple regression as a general data-analytic system. *Psychological bulletin*, *70*(6p1), 426.

Collins, A. M., & Loftus, E. F. (1975). A spreading-activation theory of semantic processing. *Psychological review*, *82*(6), 407.

Consortium, B. (2007). British national corpus. *Oxford Text Archive Core Collection.*

Dewaele, J.-M. (1998). Lexical inventions: French interlanguage as L2 versus L3. *Applied Linguistics*, *19*(4), 471-490.

Dimigen, O., Sommer, W., Hohlfeld, A., Jacobs, A. M., & Kliegl, R. (2011). Coregistration of eye movements and EEG in natural reading: analyses and review. *Journal of Experimental Psychology: General, 140*(4), 552.

Dogil, G., Frese, I., Haider, H., Röhm, D., & Wokurek, W. (2004). Where and how does grammatically geared processing take place—and why is Broca's area often involved. A coordinated fMRI/ERBP study of language processing. *Brain and Language*, *89*(2), 337-345.

Dong, J., Li, A., Chen, C., Qu, J., Jiang, N., Sun, Y., Hu, L., & Mei, L. (2021). Language distance in orthographic transparency affects cross-language pattern similarity between native and non-native languages. *Hum Brain Mapp*, *42*(4), 893-907. https://doi.org/10.1002/hbm.25266

Dong, Q. (2011). *Error Analysis of Chinese Compound Words from Vietnamese Learners of L2 Chinese* [in Chinese] (Master's thesis). Yunnan Normal University.

Du, Y., Hu, W., Fang, Z., & Zhang, J. X. (2013). Electrophysiological correlates of morphological processing in Chinese compound word recognition. *Front Hum Neurosci*, *7*, 601. https://doi.org/10.3389/fnhum.2013.00601

Eickhoff, S. B., Bzdok, D., Laird, A. R., Kurth, F., & Fox, P. T. (2012). Activation likelihood estimation meta-analysis revisited. *Neuroimage*, *59*(3), 2349-2361.

Fay, D., & Cutler, A. (1977). Malapropisms and the structure of the mental lexicon. *Linguistic inquiry*, *8*(3), 505-520.

Feng, S., Qi, R., Yang, J., Yu, A., & Yang, Y. (2020). Neural correlates for nouns and verbs in phrases during syntactic and semantic processing: An fMRI study. *Journal of Neurolinguistics*, *53*. https://doi.org/10.1016/j.jneuroling.2019.100860

Frye, R. E., Rezaie, R., & Papanicolaou, A. C. (2009). Functional neuroimaging of language using magnetoencephalography. *Physics of life reviews*, *6*(1), 1-10.

Fu, Y., Yang, J., Bermúdez-Margaretto, B., Wang, H., Cordón, D. E. J., & Domínguez, A. (2025). From eye to cortex: Tracing the neurocognitive dynamics of bilingual novel word acquisition. *Brain and Language*, *270*, 105632.

Gu, J., Li, X., & Liversedge, S. P. (2015). Character order processing in Chinese reading. *J Exp Psychol Hum Percept Perform*, *41*(1), 127-137. https://doi.org/10.1037/a0038639

Hauk, O., Davis, M. H., Ford, M., Pulvermuller, F., & Marslen-Wilson, W. D. (2006). The time course of visual word recognition as revealed by linear regression analysis of ERP data. *Neuroimage*, *30*(4), 1383-1400. https://doi.org/10.1016/j.neuroimage.2005.11.048

Haxby, J. V., Connolly, A. C., & Guntupalli, J. S. (2014). Decoding neural representational spaces using multivariate pattern analysis. *Annual review of neuroscience*, *37*(1), 435-456.

Hinke, R. M., Hu, X., Stillman, A. E., Kim, S.-G., Merkle, H., Salmi, R., & Ugurbil, K. (1993). Functional magnetic resonance imaging of Broca's area during internal speech. *Neuroreport 4(6), 675-678.*

Hoffman, P., Lambon Ralph, M. A., & Rogers, T. T. (2013). Semantic diversity: A measure of semantic ambiguity based on variability in the contextual usage of words. *Behavior research methods*, *45*(3), 718-730.

Hong, X., Huang, W., & Liu, H. (2023). The structural complexity of Chinese words and its relationship with word frequency. *Journal of Quantitative Linguistics*, *30*(3-4), 231-256.

Hou, Z. (2019). *An Analysis of Lexical Coinage Errors in Chinese by Thai Learners Based on the HSK Corpus and Its Pedagogical Implications* [in Chinese] (Master's thesis). Hebei Normal University.

Hsu, C.-H., Lee, C.-Y., & Marantz, A. (2011). Effects of visual complexity and sublexical information in the occipitotemporal cortex in the reading of Chinese phonograms: A single-trial analysis with MEG. *Brain and Language*, *117*(1), 1-11.

Hsu, C.-H., Lee, C.-Y., & Tzeng, O. J.-L. (2014). Early MEG markers for reading Chinese phonograms: Evidence from radical combinability and consistency effects. *Brain and Language*, *139*, 1-9.

Hsu, C.-H., Pylkkänen, L., & Lee, C.-Y. (2019). Effects of morphological complexity in left temporal cortex: An MEG study of reading Chinese disyllabic words. *Journal of Neurolinguistics*, *49*, 168-177. https://doi.org/10.1016/j.jneuroling.2018.06.004

Hsu, C. H., Tsai, J. L., Lee, C. Y., & Tzeng, O. J. (2009). Orthographic combinability and phonological consistency effects in reading Chinese phonograms: an event-related potential study. *Brain Lang*, *108*(1), 56-66. https://doi.org/10.1016/j.bandl.2008.09.002

Hu, Z., Zhang, J., Couto, T. A., Xu, S., Luan, P., & Yuan, Z. (2018). Optical Mapping of Brain Activation and Connectivity in Occipitotemporal Cortex During Chinese Character Recognition. *Brain Topogr*, *31*(6), 1014-1028. https://doi.org/10.1007/s10548-018-0650-y

Huang, C.-R., Chen, C.-R., & Shen, C. C. (2002). The nature of categorical ambiguity and its implications for language processing: A corpus-based study of Mandarin Chinese. *Sentence processing in East Asian languages*, 53-83.

Hung, Y.-h., Hung, D. L., Tzeng, O. J.-L., & Wu, D. H. (2014). Tracking the temporal dynamics of the processing of phonetic and semantic radicals in Chinese character recognition by MEG. *Journal of Neurolinguistics*, *29*, 42-65.

Ip, K. I., Marks, R. A., Hsu, L. S., Desai, N., Kuan, J. L., Tardif, T., & Kovelman, L. (2019). Morphological processing in Chinese engages left temporal regions. *Brain Lang*, *199*, 104696. https://doi.org/10.1016/j.bandl.2019.104696

Jin, H., Wang, Q., Yang, Y.-F., Zhang, H., Gao, M., Jin, S., Chen, Y., Xu, T., Zheng, Y.-R., & Chen, J. (2023). The Chinese Open Science Network (COSN): Building an open science community from scratch. *Advances in Methods and Practices in Psychological Science*, *6*(1), 25152459221144986.

Keuleers, E., & Balota, D. A. (2015). Megastudies, crowdsourcing, and large datasets in psycholinguistics: An overview of recent developments. *Quarterly Journal of Experimental Psychology*, *68*(8), 1457-1468.

Klein, D., Zatorre, R. J., Milner, B., & Zhao, V. (2001). A cross-linguistic PET study of tone perception in Mandarin Chinese and English speakers. *Neuroimage*, *13*(4), 646-653.

Koda, K., Coady, J., & Huckin, T. (1997). Orthographic knowledge in L2 lexical processing. *Second language vocabulary acquisition*, 35-52.

Kovelman, I., Shalinsky, M. H., White, K. S., Schmitt, S. N., Berens, M. S., Paymer, N., & Petitto, L.-A. (2009). Dual language use in sign-speech bimodal bilinguals: fNIRS brain-imaging evidence. *Brain and Language*, *109*(2-3), 112-123.

Ku, Y.-M., & Anderson, R. C. (2003). Development of morphological awareness in Chinese and English. *Reading and Writing*, *16*(5), 399-422.

Kuo, W. J., Yeh, T. C., Lee, J. R., Chen, L. F., Lee, P. L., Chen, S. S., Ho, L. T., Hung, D. L., Tzeng, O. J., & Hsieh, J. C. (2004). Orthographic and phonological processing of Chinese characters: an fMRI study. *Neuroimage*, *21*(4), 1721-1731. https://doi.org/10.1016/j.neuroimage.2003.12.007

Lai, C. S., Fisher, S. E., Hurst, J. A., Vargha-Khadem, F., & Monaco, A. P. (2001). A forkhead-domain gene is mutated in a severe speech and language disorder. *Nature*, *413*(6855), 519-523.

Laufer, B. (1988). The concept of 'synforms'(similar lexical forms) in vocabulary acquisition. *Language and education*, *2*(2), 113-132.

Law, S.-P., Yum, Y.-N., & Wing-Lam Cheung, G. (2017). Neural correlates of concreteness effect in semantic processing of single Chinese characters using mixed-effects modeling. *Journal of Neurolinguistics*, *44*, 223-238. https://doi.org/10.1016/j.jneuroling.2017.07.001

Lee, C.-Y., Hsu, C.-H., Chang, Y.-N., Chen, W.-F., & Chao, P.-C. (2015). The feedback consistency effect in Chinese character recognition: Evidence from a psycholinguistic norm. *Language and Linguistics*, *16*(4), 535-554.

Lee, C. Y., Tsai, J. L., Kuo, W. J., Yeh, T. C., Wu, Y. T., Ho, L. T., Hung, D. L., Tzeng, O. J., & Hsieh, J. C. (2004). Neuronal correlates of consistency and frequency effects on Chinese character naming: an event-related fMRI study. *Neuroimage*, *23*(4), 1235-1245. https://doi.org/10.1016/j.neuroimage.2004.07.064

Levelt, W. J. (1989). *Speaking: From intention to articulation*. MIT Press.

Li, B. (2013). An Analysis of Lexical Coinage Errors by Japanese Learners of Chinese: An Interlanguage Corpus-Based Study [in Chinese]. *Yuwen Xuekan* (04), 17-19+22.

Li, B., Ding, Z., De Deyne, S., & Cai, Q. (2024a). A large-scale database of Mandarin Chinese word associations from the Small World of Words Project. *Behav Res Methods*, *57*(1), 34. https://doi.org/10.3758/s13428-024-02513-1

Li, G., Cheung, R. T., Gao, J. H., Lee, T. M., Tan, L. H., Fox, P. T., Jack Jr, C. R., & Yang, E. S. (2006). Cognitive processing in Chinese literate and illiterate subjects: An fMRI study. *Human brain mapping*, *27*(2), 144-152.

Li, L., Ji, Y., Wang, J., & Paterson, K. B. (2025). The Impact of Color Cues on Word Segmentation by L2 Chinese Readers: Evidence from Eye Movements. *Behavioral Sciences*, *15*(7), 904.

Li, L., Yang, Y., Song, M., Fang, S., Zhang, M., Chen, Q., & Cai, Q. (2023). CCLOWW: A grade-level Chinese children's lexicon of written words. *Behav Res Methods*, *55*(4), 1874-1889. https://doi.org/10.3758/s13428-022-01890-9

Li, L., Zhao, W., Song, M., Wang, J., & Cai, Q. (2024b). CCLOOW: Chinese children's lexicon of oral words. *Behav Res Methods*, *56*(2), 846-859. https://doi.org/10.3758/s13428-023-02077-6

Li, X., Liu, P., & Rayner, K. (2011). Eye movement guidance in Chinese reading: Is there a preferred viewing location? *Vision research*, *51*(10), 1146-1156.

Lin, Y.-C., & Lin, P.-Y. (2020). Reading minds in motion: Mouse tracking reveals transposed-character effects in Chinese compound word recognition. *Applied Psycholinguistics*, *41*(4), 727-751. https://doi.org/10.1017/s0142716420000181

Lin, Y.-C., Lin, P.-Y., & Yeh, L.-H. (2023). Syllable or phoneme? A mouse-tracking investigation of phonological units in Mandarin Chinese and English spoken word recognition. *Journal of Experimental Psychology: Learning, Memory, and Cognition*, *49*(1), 130.

Linck, J. A., & Cunnings, I. (2015). The utility and application of mixed-effects models in second language research. *Language Learning*, *65*(S1), 185-207.

Liu, Y., & Perfetti, C. A. (2003). The time course of brain activity in reading English and Chinese: an ERP study of Chinese bilinguals. *Hum Brain Mapp*, *18*(3), 167-175. https://doi.org/10.1002/hbm.10090

Liu, Y., Perfetti, C. A., & Hart, L. (2003). ERP evidence for the time course of graphic, phonological, and semantic information in Chinese meaning and pronunciation decisions. *J Exp Psychol Learn Mem Cogn*, *29*(6), 1231-1247. https://doi.org/10.1037/0278-7393.29.6.1231

Liu, Y., Shu, H., & Li, P. (2007). Word naming and psycholinguistic norms: Chinese. *Behavior research methods*, *39*(2), 192-198.

Liu, Y., & Xiao, F. (2024). Effects of Lexical Properties in L2 Chinese Compound Processing: A Multivariate Approach. *Journal of psycholinguistic research*, *53*(4), 49.

Lu, C.-M., Zhang, Y.-J., Biswal, B. B., Zang, Y.-F., Peng, D.-L., & Zhu, C.-Z. (2010). Use of fNIRS to assess resting state functional connectivity. *Journal of neuroscience methods*, *186*(2), 242-249.

Luck, S. J. (2014). *An introduction to the event-related potential technique*. MIT press.

Luo, Y., Tan, D., & Yan, M. (2023). Morphological structure influences saccade generation in Chinese reading. *Reading and Writing*, *36*(5), 1339-1355.

Ma, G., Li, X., & Pollatsek, A. (2015). There is no relationship between preferred viewing location and word segmentation in Chinese reading. *Visual Cognition*, *23*(3), 399-414.

Ma, G., Li, Z., Xu, F., & Li, X. (2019). The modulation of eye movement control by word length in reading Chinese. *Q J Exp Psychol (Hove)*, *72*(7), 1620-1631. https://doi.org/10.1177/1747021818799994

Mackey, A., & Porte, G. (2012). Why (or why not), when and how to replicate research. *Replication research in applied linguistics*, *2146*, 21-46.

Marsden, E., Morgan-Short, K., Thompson, S., & Abugaber, D. (2018). Replication in second language research: Narrative and systematic reviews and recommendations for the field. *Language Learning*, *68*(2), 321-391.

Martinez, G., Conde, J., Reviriego, P., & Brysbaert, M. (2024). AI-generated estimates of familiarity, concreteness, valence and arousal for over 100,000 Spanish words. *Preprint*. https://doi.org/10.13140/RG.2.2.21597.63201

Martínez, G., Molero, J. D., González, S., Conde, J., Brysbaert, M., & Reviriego, P. (2024). Using large language models to estimate features of multi-word expressions: Concreteness, valence, arousal. *Behavior research methods*, *57*(1), 5.

Maxwell, S. E., & Delaney, H. D. (1993). Bivariate median splits and spurious statistical significance. *Psychological bulletin*, *113*(1), 181.

McCarthy, G., Blamire, A. M., Rothman, D. L., Gruetter, R., & Shulman, R. G. (1993). Echo-planar magnetic resonance imaging studies of frontal cortex activation during word generation in humans. *Proceedings of the National Academy of Sciences*, *90*(11), 4952-4956.

McClelland, J. L., & Rumelhart, D. E. (1981). An interactive activation model of context effects in letter perception: I. An account of basic findings. *Psychological review*, *88*(5), 375.

McEnery, A., & Xiao, Z. (2004). The Lancaster Corpus of Mandarin Chinese: A corpus for monolingual and contrastive language study. *Religion*, *17*, 3-4.

Meyer, D. E., & Schvaneveldt, R. W. (1971). Facilitation in recognizing pairs of words: evidence of a dependence between retrieval operations. *Journal of experimental psychology*, *90*(2), 227.

Mou, X., He, C., Tan, L., Yu, J., Liang, H., Zhang, J., Tian, Y., Yang, Y.-F., Xu, T., & Wang, Q. (2024). Chinese EEG: A Chinese linguistic corpora EEG dataset for semantic alignment and neural decoding. *Scientific Data*, *11*(1), 550.

Murphy, K. A., Jogia, J., & Talcott, J. B. (2019). On the neural basis of word reading: A meta-analysis of fMRI evidence using activation likelihood estimation. *Journal of Neurolinguistics*, *49*, 71-83. https://doi.org/10.1016/j.jneuroling.2018.08.005

Myers, J. (2006). Processing Chinese compounds: A survey of the literature. In G. Libben & G. Jarema (Eds.), *The representation and processing of compound words* (pp. 169-196).

Notaro, G. M., & Diamond, S. G. (2018). Simultaneous EEG, eye-tracking, behavioral, and screen-capture data during online German language learning. *Data in brief*, *21*, 1937-1943.

Odlin, T. (2003). Cross-linguistic influence. *The handbook of second language acquisition*, 436-486.

Palmberg, R. (1987). Patterns of Vocabulary Development in Foreign-Language Learners1. *Studies in Second Language Acquisition*, *9*(2), 201-219.

Palumbo, R. V., Marraccini, M. E., Weyandt, L. L., Wilder-Smith, O., McGee, H. A., Liu, S., & Goodwin, M. S. (2017). Interpersonal Autonomic Physiology: A Systematic Review of the Literature. *Pers Soc Psychol Rev*, *21*(2), 99-141. https://doi.org/10.1177/1088868316628405

Pan, J., Wang, A., McBride, C., Cho, J.-R., & Yan, M. (2023). Online assessment of parafoveal morphological processing/awareness during reading among Chinese and Korean adults. *Scientific Studies of Reading*, *27*(3), 232-252.

Parker, R., Graff, D., Chen, K., Kong, J., & Maeda, K. (2011). *Chinese Gigaword Fifth Edition LDC2011T13*.

Paulesu, E., McCrory, E., Fazio, F., Menoncello, L., Brunswick, N., Cappa, S. F., Cotelli, M., Cossu, G., Corte, F., & Lorusso, M. (2000). A cultural effect on brain function. *Nature neuroscience*, *3*(1), 91-96.

Peng, D., Li, Y., & Liu, Z. (1994). Identification of the Chinese Two Character Word Under Repetition Priming Condition [in Chinese]. *ACTA Psychological Sinica*, *26*, 393–400.

Perfetti, C., Cao, F., & Booth, J. (2013). Specialization and Universals in the Development of Reading Skill: How Chinese Research Informs a Universal Science of Reading. *Sci Stud Read*, *17*(1), 5-21. https://doi.org/10.1080/10888438.2012.689786

Pinheiro, E. D., Sato, J. R., Junior, R. d. S. S., Barreto, C., & Oku, A. Y. A. (2024). Eye-tracker and fNIRS: Using neuroscientific tools to assess the learning experience during children's educational robotics activities. *Trends in Neuroscience and Education*, *36*, 100234.

Plaut, D. C., McClelland, J. L., Seidenberg, M. S., & Patterson, K. (2020). Understanding normal and impaired word reading: Computational principles in quasi-regular domains. In *Connectionist psychology* (pp. 367-454). Psychology Press.

Plonsky, L. (2013). Study quality in SLA: An assessment of designs, analyses, and reporting practices in quantitative L2 research. *Studies in Second Language Acquisition*, *35*(4), 655-687.

Plonsky, L., & Oswald, F. L. (2017). Multiple regression as a flexible alternative to ANOVA in L2 research. *Studies in Second Language Acquisition*, *39*(3), 579-592.

Price, C. J. (2012). A review and synthesis of the first 20 years of PET and fMRI studies of heard speech, spoken language and reading. *Neuroimage*, *62*(2), 816-847.

Pylkkänen, L., Llinás, R., & Murphy, G. L. (2006). The representation of polysemy: MEG evidence. *Journal of Cognitive Neuroscience*, *18*(1), 97-109.

Room, A. (2013). *Dictionary of confusable words*. Routledge.

Sandra, D. (1990). On the representation and processing of compound words: Automatic access to constituent morphemes does not occur. *The Quarterly Journal of Experimental Psychology Section A*, *42*(3), 529-567.

Seidenberg, M. S., & McClelland, J. L. (1989). A distributed, developmental model of word recognition and naming. *Psychological review*, *96*(4), 523.

Sejnowski, T. J., Churchland, P. S., & Movshon, J. A. (2014). Putting big data to good use in neuroscience. *Nature neuroscience*, *17*(11), 1440-1441. https://doi.org/10.1038/nn.3839

Shen, D.-L., Bai, X.-J., Zang, C.-L., Yan, G.-L., Feng, B.-C., & Fan, X.-H. (2010). Effect of word segmentation on beginners' reading: Evidence from eye movements. *Acta Psychologica Sinica*.

Shen, D., Liversedge, S. P., Tian, J., Zang, C., Cui, L., Bai, X., Yan, G., & Rayner, K. (2012). Eye movements of second language learners when reading spaced and unspaced Chinese text. *Journal of Experimental Psychology: Applied*, *18*(2), 192.

Shen, W., Li, Z., & Tong, X. (2018). Time course of the second morpheme processing during spoken disyllabic compound word recognition in Chinese. *Journal of Speech, Language, and Hearing Research*, *61*(11), 2796-2803.

Shu, H., McBride-Chang, C., Wu, S., & Liu, H. (2006). Understanding chinese developmental dyslexia: Morphological awareness as a core cognitive construct. *Journal of Educational Psychology*, *98*(1), 122-133. https://doi.org/10.1037/0022-0663.98.1.122

Siegelman, N., Schroeder, S., Acartürk, C., Ahn, H.-D., Alexeeva, S., Amenta, S., Bertram, R., Bonandrini, R., Brysbaert, M., & Chernova, D. (2022). Expanding horizons of cross-linguistic research on reading: The Multilingual Eye-movement Corpus (MECO). *Behavior research methods*, *54*(6), 2843-2863.

Soltanlou, M., Artemenko, C., Ehlis, A. C., Huber, S., Fallgatter, A. J., Dresler, T., & Nuerk, H. C. (2018). Reduction but no shift in brain activation after arithmetic learning in children: A simultaneous fNIRS-EEG study. *Sci Rep*, *8*(1), 1707. https://doi.org/10.1038/s41598-018-20007-x

Song, D., & Li, D. (2021). Psycholinguistic Norms for 3,783 Two-Character Words in Simplified Chinese. *Sage Open*, *11*(4). https://doi.org/10.1177/21582440211054495

Steinbeis, N., & Koelsch, S. (2008). Comparing the processing of music and language meaning using EEG and fMRI provides evidence for similar and distinct neural representations. *PLoS One*, *3*(5), e2226.

Su, X., Gu, L., Hu, R., Feng, X., Li, A., Yang, J., & Mei, L. (2025). Interhemispheric functional connectivity of the ventral occipitotemporal cortex supports Chinese reading. *Brain and Language*, *270*, 105634.

Su, Y., Li, Y., & Li, H. (2022). Imageability ratings for 10,426 Chinese two-character words and their contribution to lexical processing. *Current Psychology*, *42*(27), 23265-23276. https://doi.org/10.1007/s12144-022-03404-4

Sugiura, L., Ojima, S., Matsuba-Kurita, H., Dan, I., Tsuzuki, D., Katura, T., & Hagiwara, H. (2011). Sound to language: different cortical processing for first and second languages in elementary school children as revealed by a large-scale study using fNIRS. *Cerebral Cortex*, *21*(10), 2374-2393.

Sun, C. C., Hendrix, P., Ma, J., & Baayen, R. H. (2018). Chinese lexical database (CLD): A large-scale lexical database for simplified Mandarin Chinese. *Behav Res Methods*, *50*(6), 2606-2629. https://doi.org/10.3758/s13428-018-1038-3

Sze, W. P., Rickard Liow, S. J., & Yap, M. J. (2014). The Chinese Lexicon Project: a repository of lexical decision behavioral responses for 2,500 Chinese characters. *Behav Res Methods*, *46*(1), 263-273. https://doi.org/10.3758/s13428-013-0355-9

Sze, W. P., Yap, M. J., & Rickard Liow, S. J. (2015). The role of lexical variables in the visual recognition of Chinese characters: A megastudy analysis. *Q J Exp Psychol (Hove)*, *68*(8), 1541-1570. https://doi.org/10.1080/17470218.2014.985234

Tan, C., & Zhang, Q. (2025). Open access dataset integrating behavioral and EEG measures in Chinese spoken word production. *Scientific Data*, *12*(1), 1348.

Tan, L. H., Laird, A. R., Li, K., & Fox, P. T. (2005a). Neuroanatomical correlates of phonological processing of Chinese characters and alphabetic words: a meta-analysis. *Hum Brain Mapp*, *25*(1), 83-91. https://doi.org/10.1002/hbm.20134

Tan, L. H., Liu, H.-L., Perfetti, C. A., Spinks, J. A., Fox, P. T., & Gao, J.-H. (2001). The neural system underlying Chinese logograph reading. *Neuroimage*, *13*(5), 836-846.

Tan, L. H., Spinks, J. A., Eden, G. F., Perfetti, C. A., & Siok, W. T. (2005b). Reading depends on writing, in Chinese. *Proceedings of the National Academy of Sciences*, *102*(24), 8781-8785.

Tan, L. H., Spinks, J. A., Gao, J. H., Liu, H. L., Perfetti, C. A., Xiong, J., Stofer, K. A., Pu, Y., Liu, Y., & Fox, P. T. (2000). Brain activation in the processing of Chinese characters and words: a functional MRI study. *Human brain mapping*, *10*(1), 16-27.

Tao, Y., Ficek, B., Wang, Z., Rapp, B., & Tsapkini, K. (2021). Selective functional network changes following tDCS-augmented language treatment in primary progressive aphasia. *Frontiers in Aging Neuroscience*, *13*, 681043.

Tardif, T. (1996). Nouns are not always learned before verbs: Evidence from Mandarin speakers' early vocabularies. *Developmental Psychology*, *32*(3), 492.

Tardif, T. Z. (1993). *Adult-to-child speech and language acquisition in Mandarin Chinese*. Yale University.

Theakston, A. L., Lieven, E. V., Pine, J. M., & Rowland, C. F. (2001). The role of performance limitations in the acquisition of verb-argument structure: An alternative account. *Journal of child language*, *28*(1), 127-152.

Tsang, Y.-K., Huang, J., Wang, S., Wang, J., & Wong, A. W.-K. (2024a). Comparing word recognition in simplified and traditional Chinese: A megastudy approach. *Quarterly Journal of Experimental Psychology*, *77*(3), 593-610.

Tsang, Y. K., Huang, J., Lui, M., Xue, M., Chan, Y. F., Wang, S., & Chen, H. C. (2018). MELD-SCH: A megastudy of lexical decision in simplified Chinese. *Behav Res Methods*, *50*(5), 1763-1777. https://doi.org/10.3758/s13428-017-0944-0

Tsang, Y. K., Yan, M., Pan, J., & Chan, M. Y. K. (2024b). A corpus of Chinese word segmentation agreement. *Behav Res Methods*, *57*(1), 25. https://doi.org/10.3758/s13428-024-02528-8

Tsang, Y. K., & Zou, Y. (2022). An ERP megastudy of Chinese word recognition. *Psychophysiology*, *59*(11), e14111. https://doi.org/10.1111/psyp.14111

Tse, C. S., Yap, M. J., & Chan, Y. L. (2024). Neighborhood in Chinese lexicon: A megastudy analysis of lexical decision and naming of two-character Chinese words. *J Exp Psychol Learn Mem Cogn*, *50*(9), 1489-1515. https://doi.org/10.1037/xlm0001357

Tse, C. S., Yap, M. J., Chan, Y. L., Sze, W. P., Shaoul, C., & Lin, D. (2017). The Chinese Lexicon Project: A megastudy of lexical decision performance for 25,000+ traditional Chinese two-character compound words. *Behav Res Methods*, *49*(4), 1503-1519. https://doi.org/10.3758/s13428-016-0810-5

Valaki, C., Maestu, F., Simos, P., Zhang, W., Fernandez, A., Amo, C., Ortiz, T., & Papanicolaou, A. (2004). Cortical organization for receptive language functions in Chinese, English, and Spanish: a cross-linguistic MEG study. *Neuropsychologia*, *42*(7), 967-979.

Vernes, S. C., Newbury, D. F., Abrahams, B. S., Winchester, L., Nicod, J., Groszer, M., Alarcón, M., Oliver, P. L., Davies, K. E., & Geschwind, D. H. (2008). A functional genetic link between distinct developmental language disorders. *New England Journal of Medicine*, *359*(22), 2337-2345.

Wang, M.-Y., Luan, P., Zhang, J., Xiang, Y.-T., Niu, H., & Yuan, Z. (2018). Concurrent mapping of brain activation from multiple subjects during social interaction by hyperscanning: a mini-review. *Quantitative imaging in medicine and surgery*, *8*(8), 819.

Wang, R. (2009). *The Development of Mental Lexical Representation and Mechanisms of Lexical Coinage in English-Speaking Learners of Chinese* [in Chinese] (Ph.D. thesis). Beijing Language and Culture University.

Wang, S., Zhang, Y., Zhang, X., Sun, J., Lin, N., Zhang, J., & Zong, C. (2022). An fmri dataset for concept representation with semantic feature annotations. *Scientific Data*, *9*(1), 721.

Wang, T., & Xu, X. (2025). Composition as Nonlinear Combination in Semantic Space: A Computational Characterization of Compound Processing. *Cogn Sci*, *49*(2), e70039. https://doi.org/10.1111/cogs.70039

Warriner, A. B., Kuperman, V., & Brysbaert, M. (2013). Norms of valence, arousal, and dominance for 13,915 English lemmas. *Behav Res Methods*, *45*(4), 1191-1207. https://doi.org/10.3758/s13428-012-0314-x

Westerlund, M., Kastner, I., Al Kaabi, M., & Pylkkänen, L. (2015). The LATL as locus of composition: MEG evidence from English and Arabic. *Brain and Language*, *141*, 124-134.

Wong, P. C., Kang, X., So, H.-C., & Choy, K. W. (2022). Contributions of common genetic variants to specific languages and to when a language is learned. *Scientific Reports*, *12*(1), 580.

Wu, B., Xie, Y., Lu, L., Cao, C., & Zhang, J. (2016). The construction of a Chinese interlanguage corpus. *2016 Conference of The Oriental Chapter of International Committee for Coordination and Standardization of Speech Databases and Assessment Techniques (O-COCOSDA).*

Wu, C. (2023). What is an Emotion-label word? Emotional prototypicality (EmoPro) rating for 1,083 Chinese emotion words and its relationships with psycholinguistic variables. *Journal of psycholinguistic research*, *52*(6), 2229-2237.

Wu, C. Y., Ho, M. H., & Chen, S. H. (2012). A meta-analysis of fMRI studies on Chinese orthographic, phonological, and semantic processing. *Neuroimage*, *63*(1), 381-391. https://doi.org/10.1016/j.neuroimage.2012.06.047

Wu, C., & Mu, X. (2024). Sensory experience ratings (SERs) for 1,130 Chinese words: relationships with other semantic and lexical psycholinguistic variables. *Linguistics Vanguard, 9*(1), 151-159.

Wu, J., Chang, J., Qiu, Y., & Joseph, D. (2020). The temporal process of visual word recognition of Chinese compound: Behavioral and ERP evidences based on homographic morphemes [in Chinese]. *Acta Psychologica Sinica*, *52*(2), 113-127.

Xing, H. (2003). A Statistical Analysis of Erroneous Compounds Produced by International Students [in Chinese]. *Chinese Teaching in the World* (04), 67-78+63-64.

Xu, S.-Y., Lu, F.-M., Wang, M.-Y., Hu, Z.-S., Zhang, J., Chen, Z.-Y., Armada-da-Silva, P. A., & Yuan, Z. (2020). Altered functional connectivity in the motor and prefrontal cortex for children with Down's syndrome: an fNIRS study. *Frontiers in human neuroscience*, *14*, 6.

Xu, W., Kolozsvári, O. B., Oostenveld, R., Leppänen, P. H. T., & Hämäläinen, J. A. (2019). Audiovisual processing of Chinese characters elicits suppression and congruency effects in MEG. *Frontiers in human neuroscience*, *13*, 18.

Xu, X., & Li, J. (2020). Concreteness/abstractness ratings for two-character Chinese words in MELD-SCH. *PLoS One*, *15*(6), e0232133. https://doi.org/10.1371/journal.pone.0232133

Xu, X., Li, J., & Chen, H. (2022). Valence and arousal ratings for 11,310 simplified Chinese words. *Behav Res Methods*, *54*(1), 26-41. https://doi.org/10.3758/s13428-021-01607-4

Xu, X., Li, J., & Guo, S. (2021). Age of acquisition ratings for 19,716 simplified Chinese words. *Behav Res Methods*, *53*(2), 558-573. https://doi.org/10.3758/s13428-020-01455-8

Yan, G., Tian, H., Bai, X., & Rayner, K. (2006). The effect of word and character frequency on the eye movements of Chinese readers. *British Journal of Psychology*, *97*(2), 259-268.

Yang, H., Lin, Q., Han, Z., Li, H., Song, L., Chen, L., He, Y., & Bi, Y. (2017). Dissociable intrinsic functional networks support noun-object and verb-action processing. *Brain and Language*, *175*, 29-41.

Yang, M., Liu, Y., Yue, Z., Yang, G., Jiang, X., Cai, Y., Zhang, Y., Yang, X., Li, D., & Chen, L. (2024a). Transcranial photobiomodulation on the left inferior frontal gyrus enhances Mandarin Chinese L1 and L2 complex sentence processing performances. *Brain and Language*, *256*, 105458.

Yang, Y., Li, L., de Deyne, S., Li, B., Wang, J., & Cai, Q. (2024b). Unraveling lexical semantics in the brain: Comparing internal, external, and hybrid language models. *Human brain mapping*, *45*(1), e26546.

Yao, Z., Wu, J., Zhang, Y., & Wang, Z. (2017). Norms of valence, arousal, concreteness, familiarity, imageability, and context availability for 1,100 Chinese words. *Behav Res Methods*, *49*(4), 1374-1385. https://doi.org/10.3758/s13428-016-0793-2

Yi, W. (2018). Statistical Sensitivity, Cognitive Aptitudes, and Processing of Collocations. *Studies in Second Language Acquisition*, *40*(4), 831-856. https://doi.org/10.1017/s0272263118000141

Yum, Y. N., Law, S. P., Su, I. F., Lau, K. Y., & Mo, K. N. (2014). An ERP study of effects of regularity and consistency in delayed naming and lexicality judgment in a logographic writing system. *Front Psychol*, *5*, 315. https://doi.org/10.3389/fpsyg.2014.00315

Zhang, G., Yao, P., Ma, G., Wang, J., Zhou, J., Huang, L., Xu, P., Chen, L., Chen, S., & Gu, J. (2022). The database of eye-movement measures on words in Chinese reading. *Scientific Data*, *9*(1), 411.

Zhang, H., Su, I.-F., Chen, F., Ng, M. L., Wang, L., & Yan, N. (2020). The time course of orthographic and semantic activation in Chinese character recognition: evidence from an ERP study. *Language, Cognition and Neuroscience*, *35*(3), 292-309.

Zhang, J. (2009). The Access Mechanism of Chinese Coordinative Compounds under Condition of Repetition Priming [in Chinese]. *Applied Linguistics*, *71*(3), 72–80.

Zhang, J. X., Fang, Z., Du, Y., Kong, L., Zhang, Q., & Xing, Q. (2012a). Centro-parietal N200: An event-related potential component specific to Chinese visual word recognition. *Chinese Science Bulletin*, *57* (13), 1516-1532. https://doi.org/10.1007/s11434-011-4932-y

Zhang, J. X., Xiao, Z., & Weng, X. (2012b). Neural evidence for direct meaning access from orthography in Chinese word reading. *Int J Psychophysiol*, *84*(3), 240-245. https://doi.org/10.1016/j.ijpsycho.2012.02.013

Zhang, P., Zhang, Q., Peng, G., Song, X., & Bai, X. (2016). The Representation of Chinese Two-character Words in Mental Lexicon: Evidence from an fNIRS Study [in Chinese]. *Psychological Science*, *39*(4), 849-855.

Zhang, Q., Wang, H., Luo, C., Zhang, J., Jin, Z., & Li, L. (2019). The neural basis of semantic cognition in Mandarin Chinese: A combined fMRI and TMS study. *Hum Brain Mapp*, *40*(18), 5412-5423. https://doi.org/10.1002/hbm.24781

Zhang, Q., Zhang, J. X., & Kong, L. (2009). An ERP study on the time course of phonological and semantic activation in Chinese word recognition. *Int J Psychophysiol*, *73*(3), 235-245. https://doi.org/10.1016/j.ijpsycho.2009.04.001

Zhang, Y. J., Lu, C. M., Biswal, B. B., Zang, Y. F., Peng, D. L., & Zhu, C. Z. (2010). Detecting resting-state functional connectivity in the language system using functional near-infrared spectroscopy. *J Biomed Opt*, *15*(4), 047003. https://doi.org/10.1117/1.3462973

Zhao, J., Li, Q. L., Wang, J. J., Yang, Y., Deng, Y., & Bi, H. Y. (2012). Neural basis of phonological processing in second language reading: an fMRI study of Chinese regularity effect. *Neuroimage*, *60*(1), 419-425. https://doi.org/10.1016/j.neuroimage.2011.12.074

Zhao, N., & Lei, L. (2025). Chipola: A Chinese Podcast Lexical Database for capturing spoken language nuances and predicting behavioral data. *Behav Res Methods*, *57*(6), 166. https://doi.org/10.3758/s13428-025-02697-0

Zhao, R., Fan, R., Liu, M., Wang, X., & Yang, J. (2017). Rethinking the function of brain regions for reading Chinese characters in a meta-analysis of fMRI studies. *Journal of Neurolinguistics*, *44*, 120-133. https://doi.org/10.1016/j.jneuroling.2017.04.001

Zhou, A., Yin, Y., Zhang, J., & Zhang, R. (2016). Does font type influence the N200 enhancement effect in Chinese word recognition? *Journal of Neurolinguistics*, *39*, 57-68. https://doi.org/10.1016/j.jneuroling.2016.03.001

Zhou, W., Kwok, V. P. Y., Su, M., Luo, J., & Tan, L. H. (2020). Children's neurodevelopment of reading is affected by China's language input system in the information era. *npj Science of Learning*, *5*(1). https://doi.org/10.1038/s41539-020-0062-0

Zhou, W., & Shu, H. (2017). A meta-analysis of functional magnetic resonance imaging studies of eye movements and visual word reading. *Brain Behav*, *7*(5), e00683. https://doi.org/10.1002/brb3.683

Zhou, W., Wang, A., Shu, H., Kliegl, R., & Yan, M. (2018). Word segmentation by alternating colors facilitates eye guidance in Chinese reading. *Memory & Cognition*, *46*(5), 729-740.

Zhu, Y., Xu, M., Lu, J., Hu, J., Kwok, V. P., Zhou, Y., Yuan, D., Wu, B., Zhang, J., & Wu, J. (2022). Distinct spatiotemporal patterns of syntactic and semantic processing in human inferior frontal gyrus. *Nature Human Behaviour*, *6*(8), 1104-1111.

Chapter 5
Morphological Units in the Chinese Mental Lexicon: Evidence from L1 and L2 Learners

5.1 Introduction

This series of empirical studies begins with a fundamental question concerning the morphological units of the mental lexicon. Morphological unit is central to how words are represented, accessed, and learned, and thus provides a critical entry point for understanding Chinese word processing. By examining both L1 and L2 learners, we aim to uncover how morphological units function as the building blocks of lexical knowledge across different populations. This dual perspective is essential not only for clarifying the cognitive basis of Chinese word recognition but also for exploring how morphological representations develop in second language acquisition.

For both native speakers and L2 learners, lexical representation, processing, and acquisition are inseparable (Jiang 2000). Vocabulary constitutes a crucial component constraining the development of learners' L2 proficiency, and vocabulary acquisition and growth are closely tied to the development of mental lexical representations. From the perspective of the mental lexicon, word recognition involves locating the corresponding lexical entry in the brain's mental "dictionary" and activating it to a threshold level (Elman 2004; Levelt 1989). Since Treisman (1960) firstly introduced the concept of the mental lexicon, the representational units of lexical knowledge have become a central topic of investigation. Empirical studies across a wide range of languages, including English (Feldman et al. 1995; Li et al. 2017), Chinese (Zhang and Peng 1992), Hebrew (Frost et al. 2005), Russian (Kazanina et al. 2008), Tagalog (Cayado et al. 2023), Italian (Bracco et al. 2025), and French (Giraudo et al. 2025), have provided evidence on how morphological units are represented in the mental lexicon.

Beyond this theoretical background, an additional motivation for the present research arises from corpus-based observations of Chinese L2 learners' written production. Analyses of learner corpora have revealed an interesting phenomenon of

F. Gao, *The When, Where, and How of Morphological Processing in the Human Brain*, https://doi.org/10.1007/978-981-95-5981-7_5

lexical coinage, where learners create novel, word-like forms that do not exist in standard dictionaries but remain interpretable to native speakers. For example, 兵人 (/bing1 ren2/, "soldier-person") instead of 军人 (/jun1 ren2/, "soldier") was found in L2 learners' production. Such coinages suggest that learners may rely on morphemic recombination rather than fully established whole-word representations in their mental lexicon. Previous analyses of Chinese interlanguage (Dong 2011; Li 2013; Xing 2003) have shown that lexical errors can be divided into construction errors (morpheme-level) and usage errors (whole-word level), echoing Levelt's (1989) view that lexical entries may be stored as words, morphemes, or even idioms. Statistical reanalyses of coinages from Sect. 4.1.3 further indicated that synonym morpheme substitutions account for the majority of learners' coinages (Wang 2009; Xing 2003), which might reflect incomplete semantic or morphological encoding. These findings point to the dynamic interplay between word- and morpheme-based representations in L2 mental lexicon, thereby motivating experimental investigation of how morphological units are represented and processed in both native and non-native speakers.

It remains unclear whether words from non-alphabetic languages such as Chinese are represented holistically or compositionally in the L2 mental lexicon. This question has rarely been addressed from a comparative perspective (L1 vs. L2) or a developmental perspective (high vs. low proficiency). Moreover, while morpheme property of boundness (bound vs. free) is known to modulate lexical representation in Chinese L1 speakers (Taft and Zhu 1995), little research has examined their role in L2 Chinese processing. To address these gaps, this chapter investigates the morphological units of lexical representation in both L1 and L2 Chinese speakers across different proficiency levels, with a particular focus on the influence of morpheme boundness.

Moreover, this research also contributes to Chinese vocabulary pedagogy. Existing proposals, such as "morpheme-based teaching" (语素教学) (Lyu 1999; Yu 2019) and the "whole word–character–systematic extension" (整词—析字—系连扩展) methods (Zhao 2012), highlighted the importance of morphemic awareness, yet remain fragmented and insufficiently supported by psycholinguistic evidence. By uncovering how lexical representational units evolve in L2 learners and diverge from those of native speakers, the present study offers empirical grounding for refining vocabulary instruction, clarifying the balance between whole-word and morpheme-based approaches, and developing stage-appropriate pedagogical strategies.

5.1.1 Morphological Units in English

Morphological units in the mental lexicon concern how words, particularly complex words, are stored and processed, either as holistic forms or through their decomposed constituents. Osgood and Hoosain (1974) investigated the recognition thresholds for English words, morphemes, trigrams, and noun phrases, and found that words held a distinctive salience in language comprehension. This salience was evidenced by the dominance of words over both morphemes and phrases in language processing. Based on such findings, **the "whole-word" position** was proposed, suggesting that each word is stored in the mental lexicon as an independent entry and accessed holistically, with constituent morphemes playing a relatively limited role. This hypothesis has also been supported by evidence from L2 English perception (Silva and Clahsen 2008), L2 Spanish recognition (Zyzik and Azevedo 2009), and L1 English word production (Aitchison 2012).

On the other hand, Taft and Forster (1975, 1976) demonstrated that both prefixed and polymorphemic words are decomposed into their constituent morphemes prior to lexical access, thereby supporting a **decomposition or morpheme-based view** of lexical representation. Within this framework, the morpheme, rather than the whole word, is regarded as the fundamental unit of storage and retrieval in the mental lexicon. A substantial body of evidence further indicates that morphological decomposition is not merely optional but often a necessary pathway for recognizing English words, applying to both native speakers (Bergman et al. 1988; Rastle et al. 2000; Taft and Forster 1976) and L2 learners (De Grauwe et al. 2014; Frost et al. 2005; Kim and Wang 2014; Li et al. 2017; Liang and Chen 2014; Şafak 2015; Uygun and Gürel 2017).

In addition to the whole-word and decomposition accounts, a **hybrid model** suggests that morphemes play an important but not obligatory role in lexical access (Marslen-Wilson et al. 1994; Sandra 1990; Zhou and Marslen-Wilson 1995). Within this dual-route framework, both holistic and decomposed representations of words are stored in the mental lexicon and can be accessed in parallel, with their relative activations shaped by lexical properties such as frequency, morpheme boundness, semantic transparency, and morphological complexity (Colé et al. 1997; Fehringer 2012). To further specify this architecture, Taft (1994) proposed the Interactive-Activation Model, in which morphemes and whole words are organized at different representational layers and interact dynamically during recognition. More recently, accumulating evidence has supported this hybrid pattern in L2 lexical processing across diverse learner populations, including Turkish-Dutch bilingual children (de Zeeuw et al. 2015), Chinese-speaking learners of English (Deng et al. 2017; Liang and Chen 2014), and Greek speakers of L2 English (Voga et al. 2014). These findings suggest that the representational unit in the mental lexicon should not be viewed as a strict dichotomy, but rather as a flexible system influenced by both linguistic properties and learner-related factors.

Among linguistic variables, frequency (high vs. low word/morpheme frequency) and semantic transparency (transparent vs. opaque) have not only been the most extensively studied (Sandra 1994; Wang and Peng 1999), but also shown to account for the greatest variance in lexical processing outcomes. High-frequency or semantically opaque words tend to be processed holistically, whereas low-frequency or transparent words are more likely to undergo morphological decomposition. Other lexical factors that influence representational units include word length, neighborhood or family size, familiarity, morphological structure (e.g., subordination vs. coordination), and morpheme boundness (Jiang 2013; Myers 2006).

At the same time, morphological units in the mental lexicon are also shaped by learner-related variables. Previous research has highlighted the roles of L1 background (Feng and Song 2004; Jiang 2013), AoA (Carroll and White 1973; Morrison and Ellis 2000), and L2 proficiency (Chen and Leung 1989; Liang and Chen 2014). These findings suggest that the interplay between holistic and morphemic representations is dynamic, varying across learners and developmental stages, and contingent on both linguistic input and cognitive resources.

Taken together, these insights from English studies indicate that lexical representation in the L2 mental lexicon is best understood not as a binary choice but as a continuum. It reflects the flexible interaction of multiple representational mechanisms and adapts to both language-specific structures and learner-related factors.

5.1.2 Morphological Units in Chinese

From a typological perspective, Chinese morphology contrasts markedly with that of alphabetic languages such as English, where affixation through prefixes and suffixes serves as the primary mechanism for allomorphs and word formation. In Chinese, by contrast, the system of inflection and derivation is relatively limited, and novel words are predominantly generated through compounding. More than 70% modern Chinese vocabulary consists of compound words, particularly two-character compounds (Tian and Baayen 2022; Zhou et al. 2009). This typological difference implies that conclusions drawn from studies of morphological units in alphabetic languages cannot be directly generalized to Chinese lexical processing. To address this issue, beginning in the 1990s, researchers initiated investigations into the processing of Chinese compound words.

Zhang and Peng (1992) were among the first to examine the decomposition mechanisms in Chinese compound word recognition, focusing on the effects of word frequency, morpheme frequency, and morphological structure. Their results showed that for coordinate compounds (e.g., 丢弃/diu1 qi4/, lose + discard, "abandon"), both the head and the final morpheme frequencies influenced reaction times to target words. By contrast, for subordinate structures (e.g., 荒野/huang1 ye3/, desolate + field, "wilderness"), only the frequency of the final morpheme affected response speed. Based on these findings, they proposed that Chinese words are accessed via their constituent characters or decomposed morphemes in the mental lexicon. This decomposition account has been supported by subsequent studies, including Peng et al. (1993) using a semantic priming paradigm, Tsang et al. (2014) with ambiguous morphemes, and Du et al. (2013) employing ERP techniques.

In contrast, evidence for the involvement of whole-word representations has been observed in both hybrid patterns (Peng et al. 1998; Wang and Peng 2000; Zhang et al. 2016; Zou et al. 2023) and whole-word-dominant accounts (Hong and Feng 2010; Tian et al. 2009). For instance, Peng et al. (1998) reported that both surface frequency and cumulative frequency affected response latencies to Chinese bimorphemic words. Based on these findings, they proposed the Inter/Intra

Connection (IIC) model, which posits that whole words and morphemes were encoded within the same representational layer of the lexicon, where they could exert either facilitative or inhibitory influences on one another. Particularly, Tian et al. (2009) and Hong and Feng (2010) found that native Chinese speakers tended to rely more heavily on a whole-word pathway than on morphemic information, particularly when processing higher-frequency disyllabic compounds compared to lower-frequency ones.

Empirical research on morphological units in L2 Chinese mental lexicon remains scarce and yields mixed findings. Based on lexical error analyses from Vietnamese learners' formal writing, Dong (2011) suggested that Chinese L2 beginners relied predominantly on whole-word strategies, whereas advanced learners increasingly engaged in morpheme-level decomposition when processing Chinese compound words. In contrast, Hong and Feng (2010), using a semantic priming paradigm, reported the reverse pattern. Their results indicated that L2 learners were more sensitive to morphemic information, while L1 speakers favored a whole-word pathway, although both whole-word and morpheme-level cues facilitated recognition across groups. For intermediate learners, no significant difference was observed between the homographeme priming effects from closely related word pairs (特征/te4 zheng1/–特点/te4 dian3/, "feature–characteristic") and more distantly related ones (照顾/zhao4 gu4/–照相/zhao4 xiang4/, "look after–take a picture"). However, among advanced learners, the priming effect of semantically distant pairs was significantly smaller than that of closely related ones, suggesting an increasing facilitation effect from word-level semantics. Taken together, these findings point to a developmental shift in L2 Chinese learners from morpheme-based to word-based processing with increased proficiency. Although such discrepancies may stem from differences in data sources and task designs, they underscore the need to examine more systematically how morphological units interact with L2 proficiency. Moreover, future work should also consider how lexical properties modulate this developmental trajectory.

Beyond frequency and semantic transparency (see the review in Sect. 3.6.6), another critical lexical factor is morpheme boundness, namely whether a morpheme can function as an independent word (free morpheme) or only appears within compounds (bound morpheme). The influence of morpheme boundness on lexical processing has been documented across multiple languages, including English (Forster and Azuma 2000; McKinnon et al. 2003), Japanese (Shimomura 1999), and Chinese (Hao 2015; Peng et al. 1993; Taft and Zhu 1995). Using character naming and lexical decision tasks, Taft and Zhu (1995) demonstrated that Chinese bound morphemes such as 足/zu2/ (meaning "foot" in the compound 足球/zu2 qiu2/, "football," but not used alone) can nonetheless be represented in the mental lexicon of native speakers. Similarly, Shimomura (1999), drawing on a partial repetition priming paradigm, found that for Japanese native speakers, bound morphemes (*nonword kanji*, e.g., 電 in 電気, "electricity") exerted greater facilitative effects in compound word recognition than free morphemes (*word kanji*, e.g., 人 in 人間, "human being").

Nevertheless, empirical evidence on morpheme boundness in L2 Chinese remains limited. Feng and Song (2004) directly investigated the role of morpheme boundness in L2 processing. Using a repetition priming paradigm with Korean and English learners of Chinese L2, they found that free morphemes elicited stronger priming effects than bound morphemes among Korean learners, whereas no significant difference was observed for English learners. However, the study was constrained by small sample sizes and a lack of reported effect sizes, leaving the impact of morpheme boundness in L2 Chinese largely inconclusive. This gap underscores the need for further research to clarify how morpheme boundness influences lexical access in L2 Chinese across different proficiency groups.

To date, there is no solid evidence regarding the morphological units in the mental lexicon for L2 Chinese learners across different proficiency levels or in comparison with L1 speakers. Moreover, the effect of bound and free morphemes on lexical representation remains unclear. Evidence from L1 Chinese has demonstrated the validity of the repetition priming paradigm in examining storage and decomposition processes in the mental lexicon (Peng et al. 1994; Wang and Peng 2000; Zhang 2009), suggesting that this method may also be applied productively to L2 Chinese. This chapter therefore addressed two central questions: (1) Do Chinese L2 learners at different proficiency levels and L1 speakers differ in their morphological units, as reflected in the relative robustness of whole-word versus morpheme-level priming effects? (2) When controlling for lexical factors such as frequency, semantic transparency, lexical structure, and morpheme family size, does morpheme boundness modulate lexical representation for L2 learners?

Using a repetition priming paradigm, we investigate the representational units of L1 Chinese speakers and L2 Chinese learners at intermediate and advanced levels, with particular attention to bound/free morpheme effects. In this paradigm, the target word is preceded by either itself (i.e., the identical whole-word prime), a constituent morpheme (i.e., the first-constitute morpheme), or a neutral symbol string (i.e., ####) serving as a baseline (Zhang 2009). A full priming pattern, where morphemes and whole words yield comparable priming effects, would suggest that both units are stored and retrieved as independent entries in the mental lexicon. By contrast, asymmetrical priming effects would indicate a predominant reliance on either morpheme or whole-word representations. Finally, we examine whether bound and free morphemes produce distinct priming effects in L1 and L2 groups, and interpret the findings within the interactive-activation framework.

5.2 Methods

5.2.1 Participants

Eighteen native Chinese-speaking undergraduates and graduates (10 males) who studied in Beijing were paid to participate. They were between 20 and 29 years old, all with Mandarin as their L1. Thirty-seven students from various L1 backgrounds who studied Chinese as a second language were also recruited and divided into intermediate and advanced groups. Intermediate learners (n = 18; 4 males; aged 18–28) included first- and second-year undergraduates majoring in Chinese. They had studied Chinese for about 1.8–3 years (mean = 1.9 years), held HSK[1]-3 or HSK-4 certification, and their L1s were Urdu (n = 12) and Nepali (n = 6). Advanced learners (n = 19; 12 males; aged 18–30) consisted of third- and fourth-year undergraduate students and postgraduate students. They had studied Chinese for over three years (mean = 6.7 years), all passed HSK-6, and came from diverse L1 backgrounds including Urdu (n = 2), Nepali (n = 1), Thai (n = 5), Russian (n = 3), English (n = 3), Arabic (n = 1), French (n = 1), Spanish (n = 1), Kyrgyz (n = 1), and Malay (n = 1). Learners' language backgrounds were surveyed using an adapted Chinese version of the Language History Questionnaire (Li et al. 2006). Participants with Japanese or Korean L1 were excluded to minimize prior knowledge of Chinese characters. All experimental materials and procedures were approved by the Ethics Committee of Beijing Language and Culture University.

5.2.2 Materials

Thirty high-frequency Chinese disyllabic compound words with the first morpheme being a free morpheme and another 30 with the first morpheme being a bound morpheme were selected from the Grade A and Grade B vocabulary lists of the *Outline of Chinese Proficiency Vocabulary and Character Levels* (hereafter "*Outline*", Hanban 1996). The classification of morpheme type was based on whether a morpheme appeared as a single-character word in the Grade A or B vocabulary lists of the *Outline* (Feng and Song 2004) and whether it could occur independently in modern Chinese. Four graduate students specializing in Chinese lexicology evaluated the classification, reaching an agreement rate of over 95%. All 60 compounds were high-frequency, semantically transparent words that participants had formally learned. In addition, a symbol string "****" was employed as an unrelated prime, and participants' reaction times to this symbol prime were used as the baseline for

[1]The Hanyu Shuiping Kaoshi (HSK) is a standardized test of Chinese language proficiency for L2 learners, ranging from Level 1 (basic) to Level 6 (advanced). In this study, intermediate learners were certified at Levels 3–4, while advanced learners had achieved the highest level, HSK-6.

assessing the effects of repetition priming (See Table 5.1). All experimental stimuli were provided in Appendix A.

To further ensure the familiarity of the experimental materials, a word familiarity scale was developed based on the Vocabulary Knowledge Scale (VKS; Paribakht and Wesche 1993). Five intermediate-level Chinese learners who did not participate in the formal test rated the familiarity of the target items. All items received mean scores above 4 on a 5-point scale, indicating high familiarity. For semantic transparency, 15 native Chinese speakers judged the semantic relatedness between the first morpheme and the compound word on a 5-point scale, where higher values indicated greater transparency. The transparency ratings for free-morpheme compounds (M = 4.2, SD = 0.70) and bound-morpheme compounds (M = 3.74, SD = 0.68) did not differ significantly, t (58) = 2.585, $p > 0.05$.

Morpheme family size was assessed by counting the number of words formed with each morpheme in the Grade A and B word lists of the *Outline* (Feng and Song 2004). No significant differences were found between free and bound morphemes in the *Outline* (free morphemes: M = 3.17, SD = 2.80; bound morphemes: M = 2.97, SD = 2.19), t (58) = 0.308, $p > 0.05$.

Among the real words, the distribution of word classes was as follows: 37 nouns, 19 verbs, and 4 adjectives. In terms of compound structure, 31 items were subordinate compounds, 22 were coordinate compounds, and 7 fell into other categories, closely matching the general distribution of compounding types in Chinese disyllabic words (Su 2016). In addition, character frequency and word frequency were matched across conditions using data from Modern Chinese Character Frequency Database and Modern Chinese Word Frequency Database (Xiao 2010). Moreover, 60 additional non-words were included as fillers. Table 5.2 summarizes the matching of character frequency, word frequency, familiarity, semantic transparency, morpheme family size, and stroke number across experimental conditions.

5.2.3 Procedure

Participants completed the experiment individually in a soundproof laboratory. Stimuli were presented visually using E-Prime 2.0. After reading the instruction, participants pressed the space bar to begin. A fixation cross ("+") was first presented at the center of the screen for 300 ms, followed by a blank screen for 250 ms. The prime stimulus then appeared for 350 ms, followed by a masking stimulus ("####")

Table 5.1 Experimental design and conditions

	Morpheme Priming			Whole-word Priming		Symbol Priming	
Morpheme Boundness	*Prime*	*Target*	*Translation*	*Prime*	*Target*	*Prime*	*Target*
Free Morpheme	爱	爱人	love-spouse	爱人	爱人	****	爱人
Bound Morpheme	欢	欢迎	joyous-welcome	欢迎	欢迎	****	欢迎

Table 5.2 Lexical characteristics of the experimental conditions

Morpheme Boundness	Priming Conditions	Frequency (per million)		Word Familiarity (1–5)	Semantic Transparency (0–5)	Morpheme Family Size	Stroke Number
		Character	Word				
Free Morpheme	Morpheme	1224 (1377)	509 (980)	4.2 (0.4)	4.2 (0.7)	3.2 (2.8)	7.9 (3.1)
	Whole-word	/	167 (226)			/	15.1 (4.0)
Bound Morpheme	Morpheme	784 (803)	/	4.3 (0.5)	3.7 (0.7)	3.0 (2.2)	7.3 (2.4)
	Whole-word	/	214 (206)			/	14.9 (3.6)

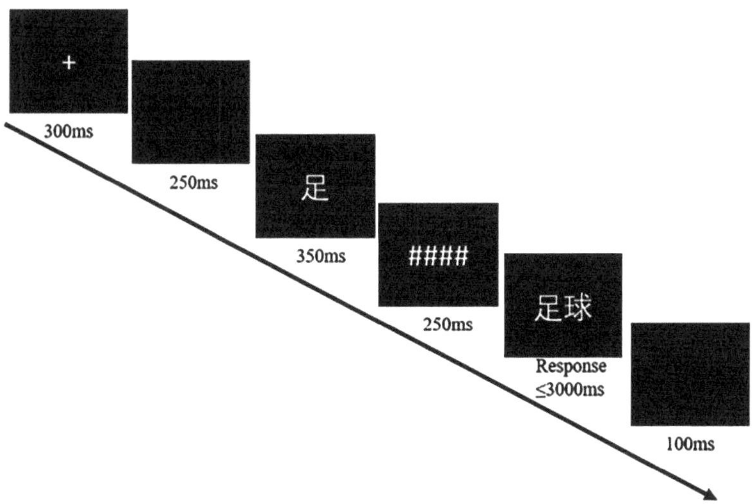

Fig. 5.1 Schematic illustration of the lexical decision task with a repetition priming paradigm

for 250 ms, after which the target word was displayed in the same font. Participants were instructed to decide whether the target was a real word or a nonword by pressing the corresponding key. If no response was made within 3000 ms, the software automatically recorded it as an error. After each response, a blank screen was shown for 100 ms before the next trial began (see Fig. 5.1).

Before the formal experiment, participants completed a practice session of 10 trials (five real-word targets and five nonword targets). Immediate feedback ("correct" or "incorrect") was provided to ensure full understanding of the task. No feedback was given during the formal experiment. Reaction times (RT) and accuracy rates (ACC) were recorded automatically by the software, and statistical analyses were conducted using IBM SPSS 21.0.

5.3 Results

Before formal statistical analyses, data from two participants were excluded due to high error rates (> 30%; one intermediate and one advanced learner), resulting in a final sample of 53 participants. Incorrect responses and RTs beyond ±2.5 SDs of each participant's mean were removed. Outliers were replaced with the participant's mean RT, accounting for less than 5% of the total data. Mean RTs and ACCs for each priming condition and participant group are presented in Table 5.3.

RT results (Fig. 5.2) revealed that the main effect of priming type was significant, F (2, 100) = 69.00, $p < 0.01$, partial $\eta^2 = 0.58$. Post hoc multiple comparisons showed significant RT differences among morpheme, whole-word, and symbol priming conditions (all $ps < 0.01$). Whole-word priming (M = 778 ms, SD = 30) produced significantly shorter RTs than morpheme priming (M = 861 ms, SD = 27), while symbol priming resulted in the longest RTs (M = 980 ms, SD = 32), indicating a robust whole-word priming effect.

The main effect of morpheme boundness was also significant, F (1, 50) = 8.07, $p = 0.006$, partial $\eta^2 = 0.14$, with free morphemes (M = 889 ms, SD = 26) eliciting longer RTs than bound morphemes (M = 857 ms, SD = 30).

A significant main effect of language proficiency was observed, F (2, 50) = 22.50, $p < 0.01$, partial $\eta^2 = .47$. Post hoc tests revealed that native speakers (M = 647 ms, SD = 48) responded faster than both intermediate learners (M = 1106 ms, SD = 49) and advanced learners (M = 865 ms, SD = 48; all $ps < .01$), and advanced learners responded faster than intermediate learners ($p < 0.01$). This demonstrates a significant effect of language proficiency on lexical decision speed.

The interaction between language proficiency and priming type was significant, F (4, 100) = 4.94, $p = 0.001$, partial $\eta^2 = 0.17$. Simple-effects analyses showed that for intermediate learners, the effect of priming type was significant, such that whole-word and morpheme priming did not differ significantly ($p = 0.08$), but both were faster than symbol priming (all $ps < 0.01$). For advanced learners, priming type had a significant effect, in which both whole-word and morpheme priming were faster than symbol priming, and whole-word priming was faster than morpheme priming (all $ps < 0.01$). For native speakers, priming type effect was significant. Whole-word

Table 5.3 Descriptive statistics of RT and ACC results across conditions and groups

		Free Morpheme			Bound Morpheme		
Group	Measures	Morpheme Priming	Whole-word Priming	Symbol Priming	Morpheme Priming	Whole-word Priming	Symbol Priming
Intermediate L2	RT (ms)	1064(48)	1068(53)	1266(56)	1075(53)	944(58)	1222(62)
	ACC (%)	92(9)	91(11)	92(9)	94(8)	90(8)	91(11)
Advanced L2	RT (ms)	869(46)	742(51)	1041(54)	825(52)	748(56)	967(61)
	ACC (%)	99(3)	99(2)	98(4)	97(5)	98(4)	97(6)
Natives	RT (ms)	666(46)	582(51)	707(54)	664(52)	584(56)	680(61)
	ACC (%)	99(2)	98(4)	99(3)	99(2)	100(0)	98(4)

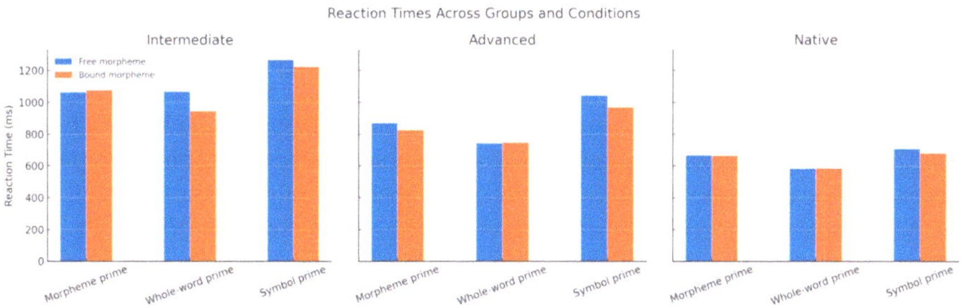

Fig. 5.2 RT results across different priming conditions among three groups

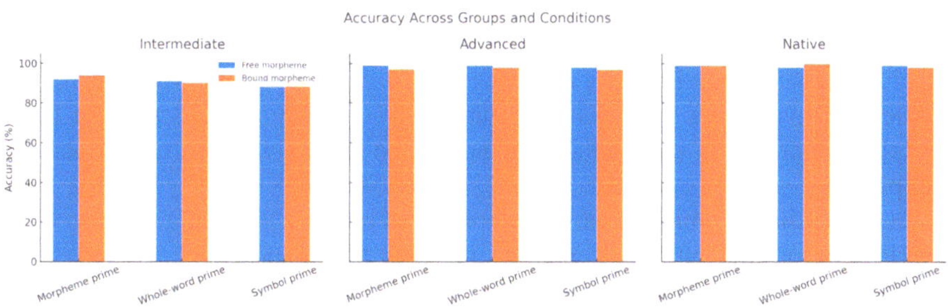

Fig. 5.3 ACC results across different priming conditions among three groups

priming was faster than morpheme priming ($p = 0.011$) and symbol priming ($p = 0.003$), whereas morpheme and symbol priming did not differ ($p = 0.743$). No other interactions reached significance.

For ACC results (Fig. 5.3), a significant main effect of language proficiency was found, $F\ (2, 50) = 16.10$, $p < 0.01$, partial $\eta^2 = 0.39$. Post hoc tests indicated that native speakers made fewer errors than both learner groups (all $ps < 0.01$), while intermediate and advanced learners did not differ ($p = 0.739$). No other main effects or interactions were significant.

5.4 Discussion

The central question addressed in this chapter concerns the basic morphological units represented in the Chinese mental lexicon. Three theoretical models have been proposed for native speakers of Chinese, i.e., the whole-word representation model, the morpheme decomposition model, and the hybrid model. Among these, the hybrid account has gained broader acceptance (Hong and Feng 2010), which assumes that both whole-word and morphemic units are simultaneously represented and operative during lexical access. Indeed, both the interactive activation model and the IIC model presupposed the coexistence of whole-word and morpheme representations. Wang and Peng (1999) further pointed out that the extent to which whole words and morphemes contribute to lexical access depends on multiple factors. In our design, we controlled for orthographic factors (stroke number), lexical properties (high-frequency words), semantic factors (high semantic transparency), and learner variables (e.g., L1 background from non-sinographic sphere). The primary goal was to assess the relative weight of whole-word and morpheme representation in lexical access, rather than to claim an absolute unit of representation.

According to the criteria for interpreting effect sizes (Cohen 2013), the main effect of priming type in this study was large (Partial $\eta^2 = 0.58$). This indicates that the cognitive distinction between morphemic and whole-word representations as separate representational units is genuine. For native Chinese speakers, the results showed that whole-word priming was significantly stronger than morpheme priming, while morpheme priming did not differ from baseline symbol priming. This indicates a dominant role of whole-word representation in lexical access. These results are consistent with Hong and Feng (2010), and support the whole-word representation hypothesis (Manelis and Tharp 1977), which suggests that the basic morphological unit in the mental lexicon may be the whole word rather than the morpheme.

In a repetition priming paradigm, whole-word priming reflects facilitation from overlapping orthography and semantics between prime and target, whereas morpheme priming reflects the combined influences of orthographic overlap and semantic relatedness between morpheme and whole word. As Zhang (2009) elaborated, recognition is facilitated when the morpheme and the target share orthography; this effect is enhanced when their meanings are related. However, when the morpheme's meaning is unrelated to the compound, inhibition can occur. If morphemic analysis were a necessary stage in word recognition, morpheme priming should not differ significantly from whole-word priming. Our finding that morpheme priming is weaker than whole-word priming thus suggests that whole-word access is more efficient, and that morphemic decomposition is not obligatory.

For intermediate-level L2 learners, whole-word and morpheme priming did not differ significantly, supporting the hybrid model. Two explanations can account for this dual-route pattern. First, whole words and morphemes may be represented at the same level in the mental lexicon. As a result, there is no difference between whole-word priming and morpheme priming, nor between free morpheme priming and bound morpheme priming. Second, following the interactive-activation account, primes first activate morphemic units, which then activate whole-word units. Lexical access is ultimately achieved through the whole-word route, although morphemic meaning can also be accessed directly. Further analysis revealed that for free morpheme priming, response times did not differ from whole-word priming, $t(16) = 0.08$, $p = 0.94$. This suggests that learners may engage in morphemic decomposition before whole-word access. However, for bound morphemes, whole-word priming was significantly faster than morpheme priming, $t(16) = 2.83$, $p = 0.012$, indicating that learners may rely more on holistic recognition when encountering compounds with bound morphemes. These findings align with the interactive-activation view, indicating that intermediate learners draw on both whole-word and analytic pathways. However, morphemic analysis appears to be a necessary stage primarily for free morphemes. Moreover, the results suggest that morphemes and whole words are represented at different levels, and that free and bound morphemes are encoded through distinct representational mechanisms.

For advanced L2 learners, as with native speakers, whole-word priming was significantly stronger than morpheme priming, suggesting a dominant role for whole-word representation. Unlike native speakers, however, advanced learners also showed faster responses to both whole-word and morpheme priming relative to baseline symbols, suggesting that morphemic representations remain partially active in their lexicon. Taken together, the results point to a developmental trajectory in L2 lexical processing. From intermediate to advanced learners, and ultimately to native-like processing, the reliance on morphemic decomposition decreases while whole-word access becomes increasingly dominant.

This developmental pattern is consistent with findings by Hong and Feng (2010), who showed that both morphemic and whole-word information contribute to recognition, but the reliance on whole-word representation increases with proficiency. Because the present study did not include beginner-level learners, our results capture only the progression from mixed representation (intermediate learners) toward whole-word dominance (advanced learners and natives). Whether beginners rely primarily on morphemic analysis remains an open question. Importantly, our study adds new evidence that morpheme type modulates this process: for intermediate learners, compounds with free morphemes tend to be accessed through morphemic decomposition, whereas those with bound morphemes are more likely processed holistically.

Our findings differ from Dong (2011), who studied Vietnamese learners and concluded that learners shift from whole-word strategies to morphemic decomposition as proficiency increases. This discrepancy may arise from methodological differences. Dong's conclusions were based on lexical errors produced at different proficiency levels, yet the classification of "whole-word errors" versus "morphemic errors" was not clearly defined. For example, advanced learners in her study showed 19.4% morphemic errors versus 13.43% whole-word errors, but no statistical tests were reported. Moreover, the corpus used was a self-built interlanguage database, with unclear reliability and operational definitions of proficiency.

Another central question in Chinese word recognition research concerns whether morpheme boundness shapes lexical access in both native speakers and L2 learners. The present study provides converging evidence that morpheme boundness exerts a robust effect on compound word processing. Taft and Zhu (1995) demonstrated in a lexical decision task with Chinese native speakers that morpheme boundness has psychological reality in the mental lexicon. Consistent with this view, our findings show that morpheme boundness significantly influences the processing of Chinese compound words. For both native speakers and learners, bound morphemes yielded a stronger priming effect than free morphemes. This pattern is also in line with Shimomura's (1999) study on Japanese compound processing using partial priming, which showed that under bound morpheme priming conditions (*nonword type*, e.g., 基–基本, "base-basic"), responses to the whole compound were faster than under free morpheme priming (*word type*, e.g., 金– 金属, "gold-metal"). However, the present results diverge from those of Feng and Song (2004). Their study showed that for Korean learners of L2 Chinese, free morphemes produced stronger priming effects than bound morphemes, whereas for European learners, the difference was not significant. According to their data, European students showed a mean priming magnitude[2] of 108 ms for free morphemes and 150 ms for bound morphemes, suggesting a numerical advantage for bound morphemes, although the main effect did not reach significance. Notably, Feng and Song did not report effect sizes for the statistical analyses. When considered together, the two studies suggest that the difference in statistical significance between free and bound morphemes may reflect variability in sample size and effect size. Importantly, because the present study recruited learners from non-sinographic sphere backgrounds, our findings parallel Feng and Song's European group, in which bound morphemes produced stronger priming effects than free morphemes.

The significant main effect of morpheme boundness in the present study further supports the interactive activation model of the mental lexicon. According to this model, free morphemes, which can function as independent words as well as bound constituents in compounds, are represented at both the lexical and morphemic levels. For instance, the character 学/xue2/ ("study/learn") is represented both as a single-word entry and as a morphemic constituent in compounds such as 学习/xue2 xi2/ ("to study/learning"). When primed by 学, both representations are activated. The single-word 学 at the lexical level competes with the target 学习, thereby inhibiting recognition, while the morphemic 学 facilitates access to the compound. These opposing influences attenuate the overall priming effect of free morphemes, resulting in slower recognition. By contrast, bound morphemes lack independent lexical entries and are represented only at the morphemic level, producing purely facilitative effects on whole-word recognition. This explains why bound morphemes yield stronger priming effects. Conversely, if both free and bound morphemes were represented at the same lexical level as compounds, no significant differences in priming would be expected.

5.5 Conclusion and Pedagogical Implications

Overall, our findings support a dual-route account of Chinese morphological processing, showing that morphological units in the mental lexicon are shaped by both linguistic properties and individual factors. While both whole-word and morphemic representations are available in L1 and L2 mental lexicon, L1 speakers and advanced learners tend to rely on more economical whole-word representations, whereas intermediate learners often display a hybrid pattern in which decomposition remains an important processing strategy. With increasing proficiency, learners gradually shift toward

[2] The priming magnitude is defined as the difference in reaction times between the control condition and the experimental condition. The larger the priming magnitude, the stronger the priming effect.

greater reliance on holistic processing, a developmental trajectory consistent with phenomena such as the stroke-number effect (Jiang et al. 2020; Su and Samuels 2010).

In direct response to the research motivation outlined in Sects. 4.1.3 and 5.1, our study provides behavioral evidence on how learners' representational units evolve, complementing the static picture provided by corpus-based analyses of lexical errors. Psycholinguistic approach in second language acquisition (SLA) research has typically focused on processing and learning, with representational questions remaining relatively underexplored (Jiang 2000). By employing a repetition priming paradigm, this study demonstrates that native and non-native speakers differ in their lexical representational units and that these differences can be explained within an interactive-activation framework. Importantly, the results reveal a dynamic continuum in learners' reliance on morphemes versus whole words, further modulated by morpheme boundness. In particular, free morphemes facilitate whole-word decomposition, especially for intermediate L2 learners.

These representational findings also bear direct relevance to pedagogical practice in teaching Chinese vocabulary. Morpheme-based instruction has long been a key topic in Chinese L2 teaching. Early advocates (e.g., Lyu 1999) emphasized that vocabulary teaching proceeds through morphemic decomposition and systematic extension, consolidating learned words and expanding new ones (Zhao 2012). Similarly, the "character-based" teaching principle (Xu 2005) and the influential view that "morpheme teaching is the foundation of L2 Chinese vocabulary instruction" (Wang and Qing 2004) have inspired extensive research on instructional materials (Yang 2011) and pedagogical practices (Chen 2006; Yu 2019). However, these studies have primarily emphasized the input of morphemes, often neglecting learners' developmental levels and lexical representation patterns. As a result, "one-size-fits-all" morpheme teaching may not always benefit learners, and may even introduce difficulties. For example, teaching free morphemes like 火/huo3/ ("fire") as productive roots may effectively support vocabulary growth (e.g., 火光/huo3 guang1/, "blaze"; 火把/huo3 ba3/, "torch"; 火山/huo3 shan1/, "volcano"), but treating bound morphemes such as 基/ji1/ ("base") in the same way is limited and potentially confusing. Moreover, there is little empirical evidence showing that morpheme extension strategies are equally effective at all proficiency levels.

Therefore, we argue for a balanced and staged integration of morpheme and whole-word teaching. For beginner and lower-intermediate learners, it may be more effective to emphasize whole-word learning, introducing morphemes selectively and cautiously. At intermediate stages, free morphemes may support lexical access, while bound morphemes are more efficiently processed as part of whole words, calling for differentiated teaching strategies. At advanced stages, learners' mental lexicons resemble those of native speakers, with whole-word dominance but residual morpheme sensitivity, suggesting that advanced instruction should prioritize whole-word storage and retrieval while also addressing morphemic structure when errors arise (e.g., overgeneralized word creation). Regardless of learners' proficiency levels, it may be beneficial to provide appropriate metalinguistic knowledge about Chinese morphology to enhance their morphological awareness.

Finally, this study has certain limitations. It did not examine the impact of different types of instructional input on representational units, nor did it address beginning learners' lexical representation. Moreover, the findings were derived solely from behavioral experiments, which provide only indirect evidence of representational units. Future research should therefore explore how classroom input interacts with representational development and whether tailored combinations of whole-word and morpheme-based teaching can promote morphological awareness and maximize learning outcomes across proficiency levels. In addition, it will be essential to employ neuroimaging methods, such as ERP, to validate the neural reality of morphological units and to uncover how they are represented in the brain.

Building on these behavioral findings, next chapter turns to the neural level of evidence. Chapter 6 examines how the brain encodes morphological constraints in Chinese compound word reading, focusing on the neural dynamics that guide lexical access. By combining cognitive theories of morphological processing with neuroimaging techniques, next chapter seeks to uncover the temporal and spatial signatures of how morphological rules are represented in the brain, thereby complementing the representational insights presented here.

References

Aitchison, J. (2012). *Words in the mind: An introduction to the mental lexicon*. John Wiley & Sons.

Bergman, M. W., Hudson, P. T., & Eling, P. A. (1988). How simple complex words can be: Morphological processing and word representations. *The Quarterly Journal of Experimental Psychology Section A*, *40*(1), 41-72.

Bracco, G., De Martino, M., & Laudanna, A. (2025). Morphological processing in written word recognition: New evidence on visual priming with stem-homographs and stem-allomorphs in Italian. *The Mental Lexicon*.

Carroll, J. B., & White, M. N. (1973). Word frequency and age of acquisition as determiners of picture-naming latency. *The Quarterly Journal of Experimental Psychology*, *25*(1), 85-95.

Cayado, D. K. T., Wray, S., & Stockall, L. (2023). Does linear position matter for morphological processing? Evidence from a Tagalog masked priming experiment. *Language, Cognition and Neuroscience*, *38*(8), 1167-1182.

Chen, H.-C., & Leung, Y.-S. (1989). Patterns of lexical processing in a nonnative language. *Journal of Experimental Psychology: Learning, Memory, and Cognition*, *15*(2), 316.

Chen, J. (2006). Some suggestions on establishing morpheme instruction [in Chinese]. *Journal of Yunnan Normal University* (6), 28-31.

Cohen, J. (2013). *Statistical power analysis for the behavioral sciences*. Routledge.

Colé, P., Segui, J., & Taft, M. (1997). Words and morphemes as units for lexical access. *Journal of Memory and Language*, *37*(3), 312-330.

De Grauwe, S., Lemhöfer, K., Willems, R. M., & Schriefers, H. (2014). L2 speakers decompose morphologically complex verbs: fMRI evidence from priming of transparent derived verbs. *Frontiers in human neuroscience*, *8*, 802.

de Zeeuw, M., Schreuder, R., & Verhoeven, L. T. (2015). Lexical processing of nominal compounds in first-and second-language learners across primary grades. *Writing Systems Research*, *7*(2), 133-156.

Deng, T., Shi, J., Bi, H., Dunlap, S., & Chen, B. (2017). The relationship between the morphological knowledge and L2 online processing of derivational words. *International Journal of Bilingualism*, *21*(4), 402-418.

Dong, Q. (2011). *Error Analysis of Chinese Compound Words from Vietnamese Learners of L2 Chinese* [in Chinese] (Master's thesis). Yunnan Normal University.

Du, Y., Hu, W., Fang, Z., & Zhang, J. X. (2013). Electrophysiological correlates of morphological processing in Chinese compound word recognition. *Front Hum Neurosci*, *7*, 601. doi:https://doi.org/10.3389/fnhum.2013.00601

Elman, J. L. (2004). An alternative view of the mental lexicon. *Trends in cognitive sciences*, *8*(7), 301-306.

Fehringer, C. (2012). The lexical representation of compound words in English: evidence from aphasia. *Language Sciences*, *34*(1), 65-75.

Feldman, L. B., Frost, R., & Pnini, T. (1995). Decomposing words into their constituent morphemes: Evidence from English and Hebrew. *Journal of Experimental Psychology: Learning, Memory, and Cognition*, *21*(4), 947.

Feng, L. P., & Song, Z. M. (2004). The Influence of the Nature and Productivity of Chinese Morphemes On the Morphological Recognition by Foreign Students [in Chinese]. *Journal of Yunnan Normal University*, *2*(6), 33–38.

Forster, K. I., & Azuma, T. (2000). Masked priming for prefixed words with bound stems: Does submit prime permit? *Language and Cognitive Processes*, *15*(4-5), 539-561.

Frost, R., Kugler, T., Deutsch, A., & Forster, K. I. (2005). Orthographic structure versus morphological structure: principles of lexical organization in a given language. *Journal of Experimental Psychology: Learning, Memory, and Cognition*, *31*(6), 1293.

Giraudo, H., Orihuela, K., & Dal Maso, S. (2025). Morphological salience effects of prefixes and suffixes embedded in French words. *The Mental Lexicon*.

Hanban. (1996). *Outline of Chinese Proficiency Vocabulary and Character Levels*[in Chinese]. Beijing Language and Culture University Press.

Hao, M. (2015). *The Development of Morpheme Awareness in Chinese Children* [in Chinese]. Beijing Language and Culture University Press.

Hong, W., & Feng, C. (2010). A Study of the Differences between CSL Learners and Chinese Native Speakers in Chinese Two-Character Compound Word Recognition [in Chinese]. *Modern Foreign Languages*, *33*(4), 387–394.

Jiang, N. (2000). Lexical representation and development in a second language. *Applied Linguistics*, *21*(1), 47-77.

Jiang, N. (2013). *Conducting reaction time research in second language studies*. Routledge.

Jiang, N., Hou, F., & Jiang, X. (2020). Analytic versus holistic recognition of Chinese words among L2 learners. *The Modern Language Journal*, *104*(3), 567-580.

Kazanina, N., Dukova-Zheleva, G., Geber, D., Kharlamov, V., & Tonciulescu, K. (2008). Decomposition into multiple morphemes during lexical access: A masked priming study of Russian nouns. *Language and Cognitive Processes*, *23*(6), 800-823.

Kim, S. Y., & Wang, M. (2014). The time-course of cross-language morphological activation in Korean-English bilinguals: Evidence from a masked priming experiment. 어학연구 *(Language Research)*.

Levelt, W. J. (1989). *Speaking: From intention to articulation*. MIT Press.

Li, B. (2013). An Analysis of Lexical Coinage Errors by Japanese Learners of Chinese: An Interlanguage Corpus-Based Study [in Chinese]. *Yuwen Xuekan* (04), 17-19+22.

Li, M., Jiang, N., & Gor, K. (2017). L1 and L2 processing of compound words: Evidence from masked priming experiments in English. *Bilingualism: Language and Cognition*, *20*(2), 384-402.

Li, P., Sepanski, S., & Zhao, X. (2006). Language history questionnaire: A web-based interface for bilingual research. *Behavior research methods*, *38*(2), 202-210.

Liang, L., & Chen, B. (2014). Processing morphologically complex words in second-language learners: the effect of proficiency. *Acta Psychol (Amst), 150*, 69-79. doi:https://doi.org/10.1016/j.actpsy.2014.04.009

Lyu, W. (1999). *A study on the grammatical system of teaching Chinese as a foreign language* [in Chinese]. Beijing Language and Culture University Press.

Manelis, L., & Tharp, D. A. (1977). The processing of affixed words. *Memory & Cognition, 5*(6), 690-695.

Marslen-Wilson, W., Tyler, L. K., Waksler, R., & Older, L. (1994). Morphology and meaning in the English mental lexicon. *Psychological review, 101*(1), 3.

McKinnon, R., Allen, M., & Osterhout, L. (2003). Morphological decomposition involving non-productive morphemes: ERP evidence. *Neuroreport, 14*(6), 883-886.

Morrison, C. M., & Ellis, A. W. (2000). Real age of acquisition effects in word naming and lexical decision. *British Journal of Psychology, 91*(2), 167-180.

Myers, J. (2006). Processing Chinese compounds: A survey of the literature. *The representation and processing of compound words*, 169-196.

Osgood, C. E., & Hoosain, R. (1974). Salience of the word as a unit in the perception of language. *Perception & Psychophysics, 15*(1), 168-192.

Paribakht, T. S., & Wesche, M. B. (1993). Reading comprehension and second language development in a comprehension-based ESL program. *TESL Canada journal*, 09-29.

Peng, D., Li, Y., & Liu, Z. (1994). Identification of the Chinese Two Character Word Under Repetition Priming Condition [in Chinese]. *ACTA Psychological Sinica, 26*, 393–400.

Peng, D., Liu, Y., & Wang, C. (1998). How is the access representation organized? The relation of polymorphemic word and their morphemes: A Chinese study. *Cognitive Analysis of Chinese Script*, 65-89.

Peng, D., Zhang, B., & Liu, L. (1993). Lexical Decomposition and Whole Word Storage of Chinese Coordination Two-Character Word [in Chinese]. In *Proceedings of the Second Afro-Asian Psychology Congress* (pp. 92–97). Peking University Press.

Rastle, K., Davis, M. H., Marslen-Wilson, W. D., & Tyler, L. K. (2000). Morphological and semantic effects in visual word recognition: A time-course study. *Language and Cognitive Processes, 15*(4-5), 507-537.

Şafak, D. F. (2015). *Morphological processing of inflected and derived words in L1 Turkish and L2 English* (Ph.D. thesis). Middle East Technical University (Turkey).

Sandra, D. (1990). On the representation and processing of compound words: Automatic access to constituent morphemes does not occur. *The Quarterly Journal of Experimental Psychology Section A, 42*(3), 529-567.

Sandra, D. (1994). The morphology of the mental lexicon: Internal word structure viewed from a psycholinguistic perspective. *Language and Cognitive Processes, 9*(3), 227-269.

Shimomura, M. (1999). Kanji lexicality effect in partial repetition priming: The relationship between Kanji word and Kanji character processing. *Brain and Language, 68*(1-2), 82-88.

Silva, R., & Clahsen, H. (2008). Morphologically complex words in L1 and L2 processing: Evidence from masked priming experiments in English. *Bilingualism: Language and Cognition, 11*(2), 245-260.

Su, B. (2016). The implicity, diversity and cognitive features of Chinese compounding [in Chinese]. *Xueshu Yanjiu* (1), 162-165.

Su, Y.-F., & Samuels, S. J. (2010). Developmental changes in character-complexity and word-length effects when reading Chinese script. *Reading and Writing, 23*(9), 1085-1108.

Taft, M. (1994). Interactive-activation as a framework for understanding morphological processing. *Language and Cognitive Processes, 9*(3), 271-294.

Taft, M., & Forster, K. I. (1975). Lexical storage and retrieval of prefixed words. *Journal of verbal learning and verbal behavior, 14*(6), 638-647.

Taft, M., & Forster, K. I. (1976). Lexical storage and retrieval of polymorphemic and polysyllabic words. *Journal of verbal learning and verbal behavior, 15*(6), 607-620.

Taft, M., & Zhu, X. (1995). The representation of bound morphemes in the lexicon: A Chinese study. *Morphological aspects of language processing*, 293-316.
Tian, H., Yan, G., & Bai, X. (2009). The access representation of Chinese two-character words in mental lexicon [in Chinese]. *Psychological Science*, *32*(6), 1302–1305.
Tian, S., & Baayen, R. H. (2022). Productivity and semantic transparency: An exploration of word formation in Mandarin Chinese. *The Mental Lexicon*, *17*(3), 458-479.
Treisman, A. M. (1960). Contextual cues in selective listening. *Quarterly Journal of Experimental Psychology*, *12*(4), 242-248.
Tsang, Y.-K., Wong, A. W.-K., Huang, J., & Chen, H.-C. (2014). Morpho-orthographic and morpho-semantic processing in word recognition and production: Evidence from ambiguous morphemes. *Language, Cognition and Neuroscience*, *29*(5), 543-560.
Uygun, S., & Gürel, A. (2017). Compound processing in second language acquisition of English. *Journal of the European Second Language Association*, *1*(1).
Voga, M., Anastassiadis-Symeonidis, A., & Giraudo, H. (2014). Does morphology play a role in L2 processing?: Two masked priming experiments with Greek speakers of ESL. *Lingvisticae Investigationes*, *37*(2), 338-352.
Wang, C., & Peng, D. (1999). The roles of surface frequencies, cumulative morpheme frequencies, and semantic transparencies in the processing of compound words. *Acta Psychologica Sinica*.
Wang, C., & Peng, D. (2000). The Access Representation of Polymorphemic Words: Decomposed or Whole? [in Chinese]. *Psychological Science*, *23*(4), 395–398.
Wang, R. (2009). *The Development of Mental Lexical Representation and Mechanisms of Lexical Coinage in English-Speaking Learners of Chinese* [in Chinese] (Ph.D. thesis). Beijing Language and Culture University.
Wang, Z., & Qing, X. (2004). Morpheme instruction as the foundation of Chinese vocabulary teaching for foreign learners [in Chinese]. *Journal of Yunnan Normal University* (5), 39-42.
Xiao, H. (2010). On the Construction and Application of Contemporary Chinese Corpus [in Chinese]. *Journal of Chinese World*, *106*, 24–29.
Xing, H. (2003). A Statistical Analysis of Erroneous Compounds Produced by International Students [in Chinese]. *Chinese Teaching in the World* (04), 67-78+63-64.
Xu, T. (2005). Character-based perspective and language research [in Chinese]. *Language Teaching and Linguistic Studies* (6), 5-15.
Yang, T. (2011). *Morpheme-based method in teaching Chinese vocabulary as a foreign language* [in Chinese] (Master's thesis). Hubei University.
Yu, Y. (2019). *A preliminary study on morpheme teaching based on international students' mental lexicon* [in Chinese] (Master's thesis). Xi'an International Studies University.
Zhang, B., & Peng, D. (1992). Decomposed storage in the Chinese lexicon. In *Advances in Psychology* (Vol. 90, pp. 131-149). Elsevier.
Zhang, J. (2009). The Access Mechanism of Chinese Coordinative Compounds under Condition of Repetition Priming [in Chinese]. *Applied Linguistics*, *71*(3), 72–80.
Zhang, P., Zhang, Q., Peng, G., Song, X., & Bai, X. (2016). The Representation of Chinese Two-character Words in Mental Lexicon: Evidence from an fNIRS Study [in Chinese]. *Psychological Science*, *39*(4), 849-855.
Zhao, J. (2012). The extraction and teaching of morphemic meaning in modern Chinese words [in Chinese]. *Chinese Teaching in the World*, *26*(3), 379–389.
Zhou, X., & Marslen-Wilson, W. (1995). Morphological structure in the Chinese mental lexicon. *Language and Cognitive Processes*, *10*(6), 545-600.
Zhou, X., Ye, Z., Cheung, H., & Chen, H.-C. (2009). Processing the Chinese language: An introduction. *Language and Cognitive Processes*, *24*(7-8), 929-946. doi:https://doi.org/10.1080/01690960903201281

Zou, Y., Tsang, Y.-K., Shum, Y.-H., & Tse, C.-Y. (2023). Full-form vs. combinatorial processing of Chinese compound words: Evidence from mismatch negativity. *International Journal of Psychophysiology*, *187*, 11-19.

Zyzik, E., & Azevedo, C. (2009). Word class distinctions in second language acquisition: An experimental study of L2 Spanish. *Studies in Second Language Acquisition*, *31*(1), 1-29.

Chapter 6
Neural Encoding of Morphological Constraints in Chinese Compound Reading

6.1 Introduction

In Chap. 5, behavioral evidence in general supported a dual-route mechanism for Chinese compound word processing. Both whole-word and morphemic representations exist in the mental lexicon, with their relative contribution shaped by lexical properties and individual factors. For native speakers and advanced L2 learners, whole-word storage appears to be the more economical and dominant strategy, whereas intermediate learners rely on both whole-word and morphemic routes in a more balanced manner.

Building on this foundation, we now turn to the neural encoding of morphological constraints in Chinese word reading. A central question is whether the brain exhibits a form of implicit sensitivity, akin to an automatic "switch", that detects the morphological legality of a string without conscious awareness. This contrasts with explicit morphological awareness and requires neuroimaging evidence to probe the time course and spatial localization of such processing. Another challenge is that morphology and semantics are often deeply intertwined, such that disentangling their respective contributions again necessitates evidence from both temporal (e.g., ERP) and spatial (e.g., fMRI/fNIRS) dimensions. To address these issues, the present chapter focuses on Chinese native speakers and Chinese disyllabic compound words to examine the neural reality of morphological constraints.

Understanding how words are represented and processed in the human brain requires uncovering the neurocognitive mechanisms of morpho-semantic processing. Morphological processing entails mental operations on a word's internal morphemic units and structure, processes that are closely intertwined with semantic interpretation (Chung et al. 2010; Ip et al. 2017). Although extensive research has examined the role of morphology in alphabetic languages (e.g., Bölte et al. 2009; Carrasco-Ortiz and Frenck-Mestre 2014; Leminen et al. 2019; Schremm et al. 2019), much less is known about how morphology is represented in the Chinese

F. Gao, *The When, Where, and How of Morphological Processing in the Human Brain*, https://doi.org/10.1007/978-981-95-5981-7_6

reading brain. Since over 70% of Chinese words are compounds composed of two or three morphemes, sub-lexical morpheme-level information is expected to be crucial for lexical access and meaning construction. Behavioral and neural evidence has indeed confirmed the psychological reality of morphemic effects and sub-lexical processing in Chinese compound reading (Huang et al. 2011a, 2011b; Zhang 1993; Zhang and Peng 1992; Zhao et al. 2017). Yet, the precise temporal dynamics and neural substrates of morphological constraints in Chinese reading remain largely unresolved.

6.1.1 Temporal Dynamics of Chinese Morphology

The temporal signature of Chinese morphological processing remains debated. One unresolved issue concerns whether sensitivity to morphological structure reflects an automatic process (e.g., Pylkkänen et al. 2004; Pylkkänen and Marantz 2003) or a more controlled, post-lexical mechanism (e.g., Allen et al. 2003). Chung et al. (2010) were among the first to investigate the time course of Chinese compound processing by presenting coordinate compounds (e.g., 花草/hua1 cao3/, flower + grass, "plants") with a short stimulus onset asynchrony (SOA) of 57 ms. Alongside the classic N400 semantic priming effect, they identified an earlier P250 effect (or P2), indexing semantic memory network activation. Crucially, the morphological structure effect was observed only in the P250 window, suggesting that morphological structure can automatically modulate semantic processing at an early stage of compound word reading. Comparable findings were later reported by Gu et al. (2012), who used a primed lexical decision paradigm manipulating both semantic relatedness and morphological structure relatedness. Their results again showed an N400 semantic effect, while the morphological priming effect emerged earlier, as an anterior P2a effect (150–180 ms). Word pairs sharing morphological structure elicited greater amplitudes than those with different structures. Together, these studies provide evidence for an early, structure-sensitive component in Chinese compound processing, distinct from later lexico-semantic effects.

By contrast, other research points to late, controlled processes (Allen et al. 2003; Newman et al. 2007). For instance, in the Chinese context, Gu and Yang (2010) manipulated word frequency (high vs. low) and morphological productivity (with subordinate compounds considered more productive than coordinate compounds, based on corpus distribution). Words of higher morphological productivity (i.e., subordinate structure) elicited significantly bigger P600 effect than those of lower morphological productivity (i.e., coordinate structure), while no early P150 or semantic N400 effect was obtained regarding distinct morphological structures. The authors interpreted the P600 effect as a late re-analysis on structural productivity or morphological structure information, which might suggest a conscious processing on Chinese compound word.

The apparent discrepancy between early and late effects may reflect differences in task demands or types of morphological structure under investigation. More importantly, these mixed findings underscore the need for finer-grained experimental designs to determine how and when morphological constraints are encoded in the time course of Chinese compound word reading.

6.1.2 Brain Localization of Chinese Morphology

While temporal evidence suggests both early and late components of Chinese morphological structure processing, much less attention has been paid to its spatial localization in the brain. Across languages, lexical access and morphological awareness are consistently associated with a semantic network involving the ventral IFG (BA45/47) and the left MTG (Arredondo et al. 2015; Bick et al. 2010; Booth et al. 2006; Chee et al. 1999; Liu et al. 2009). To date, only a limited number of studies have examined how this network supports Chinese morphology across different modalities.

In particular, Liu et al. (2013) compared children with reading disability (RD, aged 11–13) to typically developing (TD) peers in a semantic relatedness judgment task. TD children showed stronger activation in the left frontal regions (BA9/47), whereas RD children did not exhibit this pattern. Using an auditory version of the same paradigm, Zou et al. (2015) reported significant activations in the left IFG, MFG, and bilateral STG during morphological judgment. More recently, Ip et al. (2019) extended these findings and argued that morphological processing is supported by a distributed network that analyzes phonological, semantic, and structural information during word reading. Their fMRI study compared tasks (morphological vs. phonological judgment) and modalities (visual vs. auditory), revealing that morphological tasks elicited stronger activations in the dorsal IFG, posterior STG, ventral IFG, and posterior MTG. Collectively, these findings suggest that Chinese morphological processing engages both frontal and temporal regions, reflecting the morpho-syllabic characteristics of Chinese. However, most prior studies either focused on auditory input or contrasted auditory with visual tasks, leaving open the question whether a similar neural pattern is robustly observed in the visual modality for adult Chinese readers. This gap motivates the present study, which aims to examine how morphological constraints are encoded in the brain during the visual processing of Chinese compound words.

To concurrently reveal the spatio-temporal brain activation patterns associated with morphological constraints during Chinese word reading, we designed a visual lexical decision task with disyllabic Chinese compound words, pseudowords, and nonwords. Interestingly, the construction of pseudowords in our design was partly inspired by L2 learners' lexical coinages (see Sect. 4.1.3), such as 兵人 /bing1 ren2/ ("soldier-person") for 军人/jun1 ren2/ ("soldier"), which highlight how learners spontaneously combine morphemes into novel but interpretable forms. In this context, we operationalized morphological constraints as sensitivity to morphological structure, reflecting the ability to resolve structural challenges in word recognition. Specifically, pseudowords provide an effective probe for morphological processing

(Bölte et al. 2009). Pseudowords such as 兵人 are semantically interpretable, but when language users attempt to map them onto established lexical entries in the mental lexicon, they encounter difficulties in structural integration. Thus, the comparison between real words and pseudowords indexes morphological sensitivity, whereas the contrast between real/pseudowords and nonwords taps into semantic analysis.

To capture the neural basis of these processes, we employed simultaneous EEG–fNIRS recordings. EEG provides millisecond-level resolution of cortical dynamics in response to stimuli (Luck 2014), whereas fNIRS localizes hemodynamic changes in ecologically flexible settings including reading (Hofmann et al. 2014; Pinti et al. 2020). The integration of these two modalities enables a comprehensive characterization of neurovascular coupling, thus paving a novel avenue for examining when, where, and how Chinese morphological constraints are encoded in the brain during compound word reading. Importantly, this multimodal approach not only helps disentangle the contributions of morphological and semantic factors, but also provides neural evidence to further test the dual-route model of Chinese compound processing proposed in Chap. 5.

6.2 Methods

6.2.1 Participants

Twenty-one native speakers of Mandarin Chinese (12 females) from the University of Macau participated in the study. Participants were between 19 and 31 years of age (M = 25.3) and were all right-handed, as determined by the Edinburgh Handedness Inventory (Oldfield 1971). All had normal or corrected-to-normal vision and reported no history of neurological or psychiatric disorders. Written informed consent was obtained from each participant prior to testing, and all procedures were approved by the Institutional Review Board of University of Macau.

6.2.2 Materials

The task employed three categories of stimuli: real words, pseudowords, and nonwords. The real-word set consisted of 45 high-frequency Chinese disyllabic compounds (frequency > 50 per million), drawn from the CNCORPUS database (Jin et al. 2005). Pseudowords were generated by substituting one morpheme of a real compound with a semantically related morpheme (Ip et al. 2019), while ensuring that the resulting form adhered to the legitimate morphological compounding patterns in Chinese, such as coordinate, subordinate, verb–resultative, subject–predicate, and verb–object structures. In contrast, nonwords were constructed by

Table 6.1 Lexical properties of real words, pseudowords, and nonwords

	Real word		Nonword		Pseudoword	
	Mean	SD	Mean	SD	Mean	SD
Morphological legality rating	/	/	2.13	2.05	20.27	2.15
Stroke number	15.91	3.71	15.13	3.7	15.78	4.23
Word frequency	105	199	/	/	/	/
Position frequency (head)	0.58	0.2	0.36	0.24	0.59	0.23
Position frequency (end)	0.58	0.18	0.31	0.22	0.61	0.18
Character frequency (head)	581	795	1743	7208	1094	1530
Character frequency (end)	924	1230	876	8560	1358	1622

[a]Frequency: occurrence per million

arbitrarily combining two characters from the same database to form pronounceable but meaningless and morphologically ill-formed strings (Leong et al. 2005), such as 寸互/cun4 hu4/, ("inch-mutual"), which yield no interpretable meaning.

To ensure the reliability of pseudoword construction, 20 raters with backgrounds in Chinese linguistics were asked to evaluate whether each string was morphologically legal and, if so, to specify its structure from the five available options. Only those pseudowords that were consistently identified as belonging to a specific morphological structure by at least 16 raters were retained in the stimulus pool. For example, the pseudoword 冷情 (冷/leng3/, "cold"; 情 /qing2/, "feeling") was derived from the real word 热情/re4 qing2/ (hot + feeling, "enthusiasm"). While 冷情 is not an attested entry in Chinese dictionaries, it conforms to compounding rules (i.e., subordinate structure) and yields an interpretable meaning, making it a suitable pseudoword candidate. In total, 45 pseudowords were selected from the validated pool. The distribution of morphological structures among pseudowords (subordination = 27, coordination = 8, verb–object = 9, verb–complement = 1, subject–predicate = 0) and real words (subordination = 21, coordination = 17, verb–object = 4, verb–complement = 2, subject–predicate = 1) closely mirrored the proportions reported in the Chinese lexicon (Su 2016).

In addition, three separate sets of 45 pseudowords, 45 nonwords, and 15 real-word fillers were evaluated by 22 experts in Chinese linguistics, who judged whether each item was morphologically legal or not. The results showed that pseudowords were judged as morphologically legal significantly more often than nonwords, $t(88) = 40.97$, $p < 0.001$ (see "Morphological legality rating" in Table 6.1). Furthermore, multiple comparisons were conducted on position frequency statistics (head and end). As shown in Table 6.1, the position frequency of nonwords was significantly lower than that of both real words and pseudowords ($ps < 0.05$), while no difference emerged between the real words and pseudowords ($ps > 0.05$). In contrast, no significant differences were found among the three conditions with respect to stroke numbers or constituent character frequency ($ps > 0.1$). All experimental stimuli were provided in Appendix B.

6.2.3 Procedure

Participants were seated comfortably in a dimly lit room, approximately 80 cm away from the computer monitor. Before the formal experiment, they completed a short practice session consisting of three real-word and three nonword trials. During both practice and testing, participants were instructed to judge whether the character string presented on the screen was a real Chinese word by pressing the corresponding labeled keys. After practice, they rested for 30 s before proceeding to the main task, which was organized into three blocks.

Each block included 15 real words, 15 pseudowords, 15 nonwords, and 5 real-word fillers. A one-minute break was given between blocks. As illustrated in Fig. 6.1, each trial began with a fixation cross (500 ms), followed by a blank screen (200 ms), target word presentation (400 ms), and a probe with two question marks (up to 3000 ms) for response. Participants needed to judge whether the previous presentation was a read word or not when they saw the question marks. This sequence ended with a red cross displayed for 12,000 ms. The extended rest period was implemented because of the slow event-related design of fNIRS, which requires sufficient time for the hemodynamic response to return to baseline between trials (Hu et al. 2018). The entire experiment lasted approximately 40–50 min.

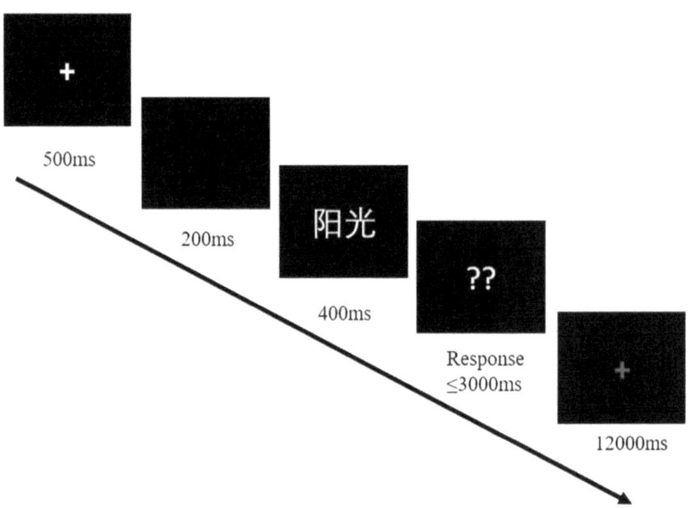

Fig. 6.1 The schematic illustration of the experimental procedure

6.2.4 EEG-fNIRS Data Acquisition and Analysis

EEG and fNIRS signals were recorded concurrently while participants performed the lexical decision task, using a fused fNIRS–EEG acquisition cap (EasyCap, Herrsching, Germany) aligned with the international 10–20 system. The fNIRS data were collected with the NIRScout system (NIRx Medizintechnik GmbH, Berlin, Germany), equipped with 8 LED sources and 8 detectors positioned over the frontal and temporal regions of the left hemisphere (Zhang et al. 2016; Zhang 2019). Light attenuation was measured at wavelengths of 760 nm and 850 nm with a sampling rate of 7.81 Hz, resulting in 22 measurement channels (Fig. 6.2) with an inter-optode distance of 30 mm.

To obtain anatomical information, the Montreal Neurological Institute (MNI) coordinates of each fNIRS channel were estimated using the ICBM-152 head model (ICBM 2009a Nonlinear Symmetric Atlas) in combination with the NIRSite 2.0 toolbox (NIRx Medizintechnik GmbH, Berlin, Germany), based on the EEG 10–20 system (Chatrian et al. 1985; Pfeifer et al. 2017). These coordinates were then imported into the NIRS-SPM toolbox (Ye et al. 2009) for spatial registration. The spatial properties of the 22 channels, including MNI coordinates, BA, anatomical labels, and overlap percentages, are provided in Appendix C.

The fNIRS data were preprocessed using nirsLAB (v2019.04; Xu et al. 2014). The raw signals were band-pass filtered between 0.01–0.2 Hz to eliminate drift and respiratory noise, with the 2-second period preceding stimulus onset used as baseline. Discontinuities and spike artifacts were removed with the software's built-in algorithms (standard deviation threshold set at 5). Hemodynamic changes in oxygenated hemoglobin (HbO), deoxygenated hemoglobin (HbR), and total hemoglobin (HbT) were subsequently calculated using the modified Beer–Lambert law (Baker et al. 2014). Because HbO has been shown to be a more sensitive measure of cortical activation (Fu et al. 2014; Lin et al. 2018), only HbO signals were analyzed for this study. To extract brain activity parameters for each lexical condition, the HbO time series were modeled using statistical parametric modeling (SPM Level 1 module in nirsLAB), with a canonical hemodynamic response function (HRF) (parameters: 6, 16, 1, 1, 6, 0, 32) applied to estimate general linear model (GLM) coefficients (beta weights).

In parallel with fNIRS, EEG data were collected concurrently to capture the fine-grained temporal dynamics of morphological processing. Recordings were obtained with 32 Ag/AgCl active electrodes (Brain Products, Munich, Germany) mounted on the EEG cap (sampling rate: 500 Hz; band-pass: 0.03–70 Hz; reference: TP9). Because several sites overlapped with the placement of fNIRS probes, six electrodes were shifted to adjacent positions (see Fig. 6.2). Electrode impedance was kept below 25 kΩ throughout data acquisition.

The raw EEG signals were processed offline using Brain Vision Analyzer 2.1 (Brain Products GmbH). Data were re-referenced to the averaged mastoids (TP9, TP10) and filtered between 1–30 Hz. To remove ocular artifacts such as blinks, independent component analysis (ICA) was applied. Trials containing artifacts

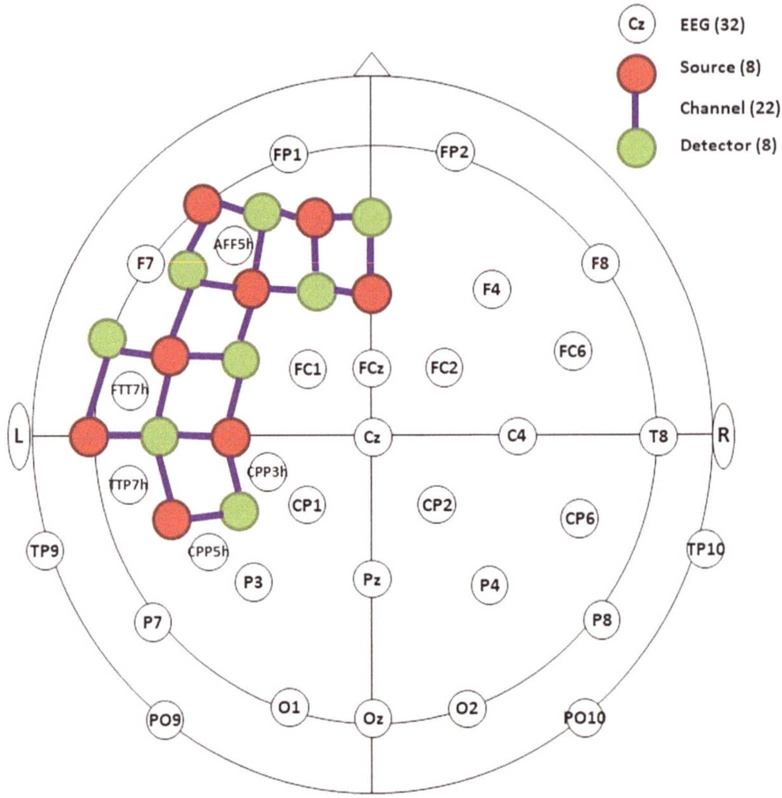

Fig. 6.2 The layout of EEG electrodes and fNIRS optodes (including sources and detectors)

exceeding ±80 μV were excluded. The data were segmented from 200 ms before to 800 ms after stimulus onset, with the 200 ms pre-stimulus period serving as baseline. For all three lexical conditions, only correctly responded trials were retained for analysis, with less than 15% of trials discarded due to artifacts.

6.3 Results

6.3.1 Behavioral Results

The ACC and RT data were respectively subjected to one-way repeated measures ANOVA with the lexical condition (real word, pseudoword, and nonword) as an independent variable. The analysis results revealed a significant main effect of lexical condition on RT, $F\,(2{,}40) = 20.23$, $p < 0.001$, partial $\eta^2 = 0.5$. Pairwise comparisons showed that RT of real words (703 ± 88 ms) was significantly shorter than

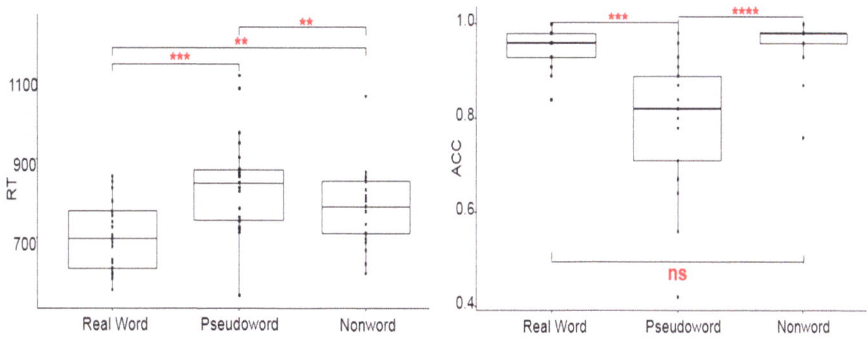

Fig. 6.3 RT (ms) and ACC across the three lexical conditions

those of pseudowords (834 ± 126 ms) and nonwords (775 ± 99 ms), while pseudowords exhibited statistically longer RT than nonwords. In addition, the ANOVA results of mean accuracy also illustrated a reliable main effect of lexical condition: $F(2,40) = 27$, $p < 0.001$, partial $\eta^2 = 0.58$. It was found that the accurate responses to real words (95.4% ± 4.2%) and nonwords (95.3% ± 5.8%) were significantly higher than that to pseudowords (79.5% ± 14.1%), whereas no significant difference in mean accuracy was detected between the nonwords and real words. RT and ACC results were visualized in Fig. 6.3. The significance of multiple comparisons was provided (**$p < 0.01$, ***$p < 0.001$, ****$p < 0.0001$).

6.3.2 *ERP Results*

EEG data from three participants were discarded due to extensive body movement and physiological noise. For all the other participants, grand-averaged ERPs were generated across all lexical stimuli. As shown in the butterfly plot (Fig. 6.4), two consecutive negative-going peaks were detected after 280 ms post stimuli onset. The time windows and regions of interest (ROIs) for the two ERP components were denoted by the negativity distribution on the scalp. Interestingly, an early ERP component N400 was detected along the fronto-central electrodes (left: FC1, midline: FCz, right: FC2) within the time window of 280–340 ms post stimuli, whereas the late N400 was identified at the centro-parietal cortex (left: CCP3h, CP1, midline: Cz, right: C4, CP2) from 340 ms to 520 ms. More importantly, previous ERP studies have demonstrated that the two ERP components including early N400 and late N400 were associated with Chinese morpho-orthographic and morpho-semantic processing, respectively (Wu et al. 2020).

Early N400. A two-way repeated measures ANOVA with Geisser-Greenhouse correction (Greenhouse and Geisser 1959) was performed on the peak of early N400 (280–340 ms), with lexical condition and laterality (left, midline, vs. right) as factors. It was found that the main effect of lexical condition was significant, *F*

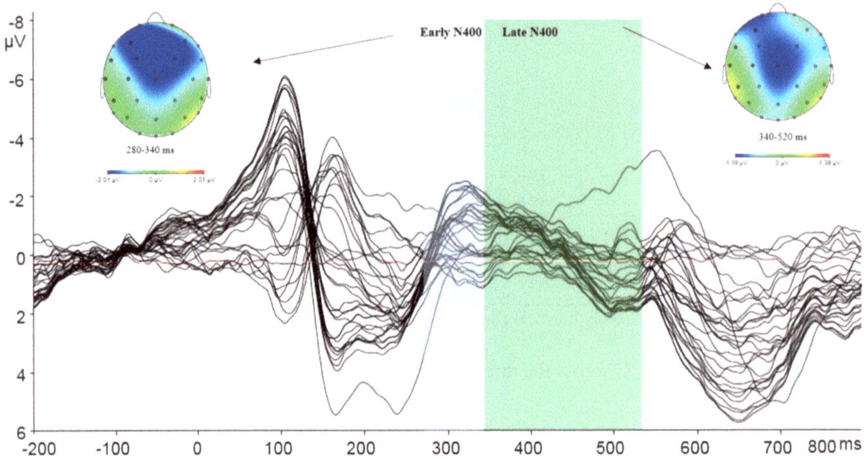

Fig. 6.4 Butterfly plot of grand-averaged ERPs of all lexical stimuli depicting averaged ERP traces for all channels across all participants

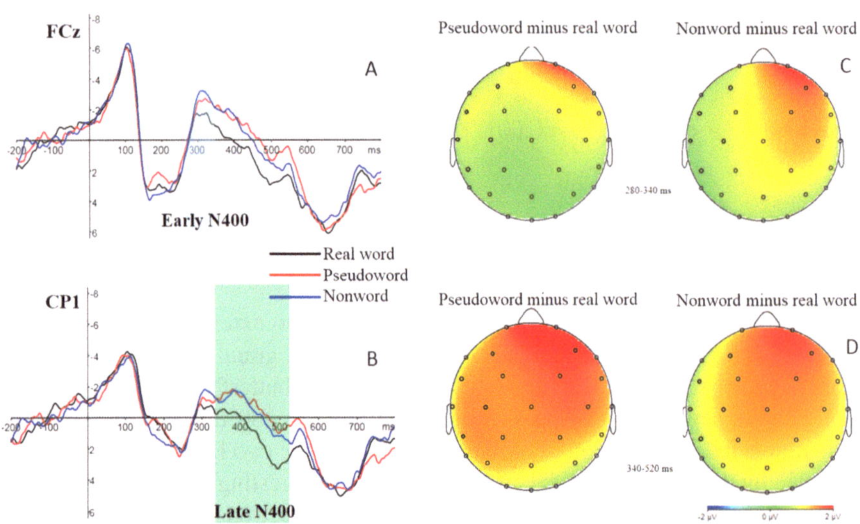

Fig. 6.5 Grand-averaged ERPs on representative electrodes and voltage maps showing early and late N400 effects

(2,34) = 3.68, $p = 0.036$, partial $\eta^2 = 0.18$. Further comparison analyses revealed that the nonword case showed larger negativity as compared to the real word case (mean peak value: −4.12 vs. -2.77 μV, $p = 0.008$), although no obvious difference was identified for the cases of nonword vs. pseudoword and real word vs. pseudoword contrast (*ps* > 0.05). Besides, no significant main effect was detected for the laterality [F (2,34) = 0.44, $p = 0.65$, partial $\eta^2 = 0.025$], or interaction between the lexical condition and laterality [F (4,68) = 1.09, $p = 0.37$, partial $\eta^2 = 0.06$] (Fig. 6.5).

Late N400. An identical repeated measures ANOVA was carried out for late N400, indicating that the main effect of lexical condition was significant, $F(2,34) = 5.4$, $p = 0.009$, partial $\eta^2 = 0.24$. Additional analyses demonstrated that the pseudoword case induced a larger N400 than real word case (−3.4 vs. −2 μV, $p = 0.006$), and the nonword case elicited a larger N400 than the real words (−3.1 vs. −2 μV, $p = 0.022$), whereas no statistical difference was revealed between the pseudowords and nonwords ($p = 0.51$). Further statistical analyses indicated that the laterality effect was also significant, $F(2,34) = 8.46$, $p = 0.001$, partial $\eta^2 = 0.33$. In particular, the midline electrodes induced a larger N400 (−3.3 μV) than the bilateral sites (left: −2.6 μV; right: −2.6 μV; $ps < 0.01$). By contrast, no interaction between the lexical condition and laterality was identified for later N400 [$F(4,68) = 1.36$, $p = 0.06$, partial $\eta^2 = 0.07$].

6.3.3 fNIRS Results

Channel Analysis. Pairwise comparisons were conducted channel by channel to inspect the difference in cortical hemodynamic response between the three lexical conditions. Interestingly, analysis results from paired-samples t tests (Table 6.2; Fig. 6.6) revealed significant differences (uncorrected $p < 0.05$) in channels 2, 9, and 12. In particular, the pseudoword case showed enhanced brain activation as compared to the real-word condition in channel 2 (including frontopolar area, IFG, and dorsolateral prefrontal cortex-DLPFC). Meanwhile, compared to the real words, the non-word case elicited greater brain activation in channels 9 and 12 (DLPFC and pars triangularis Broca's area). However, only channel 9 exhibited a significant difference in brain activation between the nonwords and real-word cases after adjusting p values for FDR (false discovery rate) (Noble 2009). In Fig. 6.6, t values are displayed according to the color bar. The solid circle denotes the channel that shows significantly enhanced activations ($p < 0.05$ FDR-adjusted), whereas the dashed circles represent unadjusted $p < 0.05$.

ROI Analysis. Further, two ROIs were extracted based on previous work (Ip et al. 2019; Zhang et al. 2016), which covered the left prefrontal area (BA 9/10, 45/46; channels 2, 7, 9, and 12) and the left temporal cortex (BA 20/21/22; channels 15, 16, and 17). One-way repeated measures ANOVA was performed for each of the two ROIs, followed by pairwise comparisons to obtain FDR-corrected p values for detecting the difference between each two conditions. A significant effect of lexical condition was found for ROI covering the left prefrontal area [$F(2,40) = 4.4$, $p = 0.018$, partial $\eta^2 = 0.18$]. Further comparisons demonstrated that the brain activation of pseudoword or nonword cases was significantly higher than that of real word condition (FDR adjusted $ps < 0.05$, see Fig. 6.7a and c). However, no significant difference between the lexical conditions was identified for ROI along the left temporal cortex, $F(2,40) = 0.08$, FDR adjusted $p = 0.9$, partial $\eta^2 = 0.18$ (Fig. 6.7b and d).

Table 6.2 fNIRS channels showing different HbO levels between the three conditions

CH#	MNI coordinates			BA	Anatomical label	Overlap	Comparisons	*t*	*df*	*p* (uncorrected)	*p* (FDR corrected)
	x	y	z								
2	−48	51	−3	10	Frontopolar area	0.4926	Pseudowords minus Real Words	2.2	20	0.044	0.131
				47	Inferior prefrontal gyrus	0.4926					
				46	Dorsolateral prefrontal cortex	0.3889					
9	−47	32	37	9	Dorsolateral prefrontal cortex	0.5875	Nonwords minus Real Words	2.7	20	0.014	0.042
				46	Dorsolateral prefrontal cortex	0.3333					
12	−57	29	13	45	pars triangularis Broca's area	0.659	Nonwords minus Real Words	2.1	20	0.048	0.145
				46	Dorsolateral prefrontal cortex	0.341					

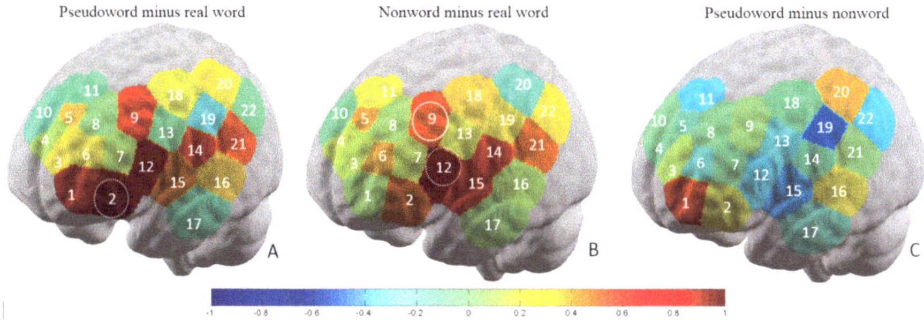

Fig. 6.6 Brain maps of *t* values of HbO signal contrast between the three lexical conditions

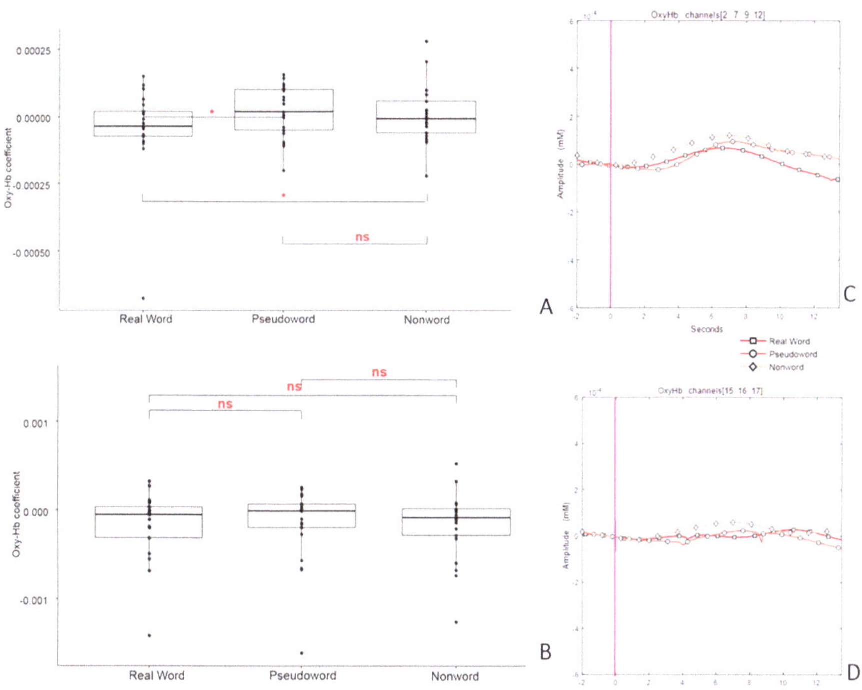

Fig. 6.7 The brain activation comparisons across three conditions along the left prefrontal region (**a** & **c**, channels 2, 7, 9, 12) and the left temporal region (**b** & **d**, channels 15, 16, 17)

6.3.4 Correlational Results

The relationship between participants' behavioral performance and their brain activation detected by ERP or fNIRS data were carefully inspected (Fig. 6.8). For brain regions detected by fNIRS data, a significantly negative correlation between the participants' reaction time to pseudo-words and HbO signal in the prefrontal area

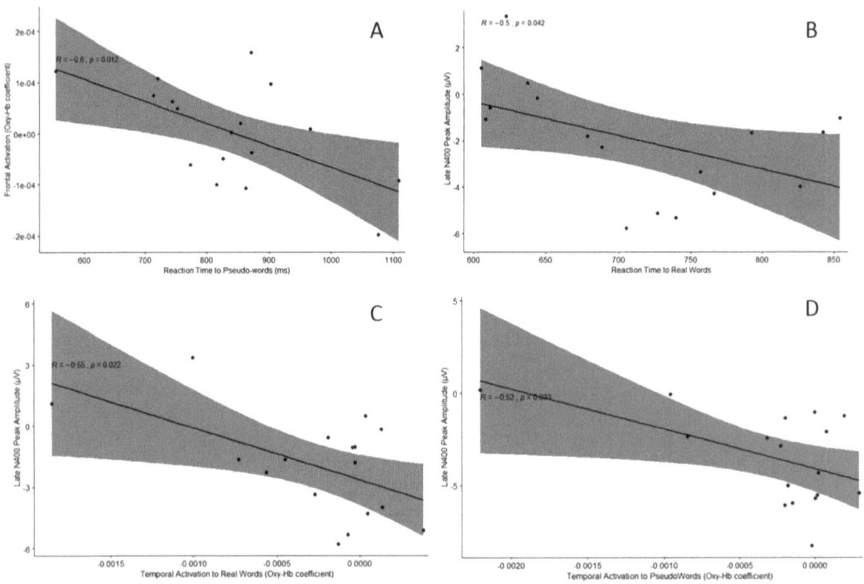

Fig. 6.8 Correlational results across behavioral, EEG, and fNIRS data

was identified (Fig. 6.8a). Meanwhile, for ERP data, the RT of real words exhibited significant relationship with late N400 at CP1 electrode. Further, the relationship between the two ERP components and HbO signals was also carefully examined. It was discovered that participants' hemodynamic responses when reading real words and pseudowords tended to be closely correlated with late N400.

6.4 Discussion

In this study, we employed a visual lexical decision paradigm combined with fused EEG–fNIRS recordings to examine the behavioral patterns, temporal dynamics, and spatial activations associated with morpho-semantic constraints in adult native Chinese readers. The behavioral results revealed clear effects of both semantic regularity and morphological constraints. RTs were significantly longer for semantically anomalous items (nonwords) than for real words, indicating a semantic processing cost. Similarly, the contrast between real words and pseudowords demonstrated readers' sensitivity to morphological legality and the additional structural parsing demands posed by pseudowords. Accuracy was higher for real words and nonwords than for pseudowords, further underscoring the latter's structural integration difficulty, which may reflect the additional demands of decomposing pseudowords during lexical access. These findings support our initial hypothesis that both morphological and semantic factors contribute to Chinese compound word recognition. The behavioral patterns are also consistent with the dual-route account outlined in Chap. 5, suggesting that both whole-word and morphemic representations are engaged during recognition, with their relative contributions modulated by lexicality constraints.

Turning to the ERP findings, distinct temporal activation patterns were observed for real words, pseudowords, and nonwords. Most notably, nonwords elicited greater negativity than real words within the 280–340 ms window, corresponding to the early N400 component. Recent accounts of the early N400 effect (280–400 ms; Wu et al. 2020) suggest that it reflects morphemic-level processing in Chinese complex word recognition. Given that nonwords are novel stimuli with extremely low or zero lexical frequency, participants may have first decomposed them into familiar constituent characters before attempting to establish a possible semantic relation between them. Thus, the enhanced negativity for nonwords relative to real words in this time window likely indexes semantic analysis at the morphemic level, consistent with a sub-lexical decomposition account (Feldman et al. 1995; Huang et al. 2011a, 2011b).

Interestingly, following the early N400 semantic modulation, a morphological legality effect emerged between 340–520 ms. Specifically, pseudowords elicited a significantly larger late N400 than real words, with the effect most pronounced along the midline. This suggests that the late N400 is particularly sensitive to morphological status in complex Chinese word recognition. Although pseudowords conform to compounding rules, they appear to demand additional cognitive resources as readers attempt to resolve structural organization when conventional semantic analysis is insufficient. For instance, with the pseudoword 冷情 ("cold + feeling"), participants can readily access the meanings of its constituent morphemes, yet encounter difficulty mapping the structure onto any existing lexical entry in the mental lexicon. Thus, the enhanced late N400 may reflect a controlled process of structural evaluation, highlighting the interaction between morphology and semantic analysis under conditions of morphological constraint (Allen et al. 2003; Gu and Yang 2010; Newman et al. 2007). Furthermore, Wu et al. (2020) demonstrated that the late N400 (430–540 ms) is associated with morpho-semantic analysis, as reflected in differential brain responses to homophonic–homographic versus heterophonic–homographic words, whereas the early N400 was linked to morpho-orthographic processing at the sub-lexical level. Based on this distinction, we infer that morphological structure processing in Chinese may rely on a controlled neural process.

Besides, the early N400 detected at 280–340 ms exhibited no morphological discrimination between the pseudowords and nonwords. By contrast, the temporal neural activation difference between the nonwords and real words was persistent across both early- and late-stage time windows (from 280–340 ms to 340–520 ms). Hence, semantic analysis at morphemic level occurred earlier till the content of stimuli were cracked. And then the constitutes were combined together to fit into the existing morphological rules. In particular, the late N400 effect for pseudoword recognition indicated that these constitutes are further processed holistically after their decomposition and re-composition. Note that a significant relationship was detected between the RT and late N400 response to real words, representing an index of cognitive resource consumption in intentional processing of post-lexical information (e.g., structural aspect). These findings collectively shed light upon the relative independence of morphological processing, which might encode a controlled process of neural resources.

Previous studies have suggested that the ERP component P2 may be linked to Chinese morphological structure processing (Chung et al. 2010; Gu et al. 2012). However, no such early effect was observed in the present study. One possible explanation lies in the methodological differences: earlier work employed a priming paradigm with very short SOA (57 ms and 200 ms), whereas our paradigm involved word reading in isolation with a longer processing window. Thus, the P2 effect reported previously may reflect the specific processing demands of priming rather than general word recognition. Future research is needed to clarify whether the P2 component is uniquely tied to morphological constraints under priming conditions.

While EEG provides high temporal precision for identifying the time course of morphological effects, fNIRS complements this by mapping the spatial patterns of brain activation through hemodynamic responses. In the present study, both channel-based and ROI-based analyses pointed to a central role of the left prefrontal cortex in Chinese morphological processing. Specifically, channel 2 (BA10/47) was highly sensitive to morphological regularity in the pseudoword versus real word contrast, whereas channels 9/12 (BA9/45/46) distinguished between semantically irregular and regular conditions. ROI analyses further revealed that morphological sensitivity (pseudowords vs. real words) and semantic evaluation (nonwords vs. real words) both engaged the left prefrontal region (Booth et al. 2006; Liu et al. 2009). These findings validate the crucial role of the left IFG (BA9/47) in Chinese morphological processing.

Importantly, our results converge with cross-linguistic evidence. For instance, English-speaking children showed significant activation in the left IFG during auditory morphological awareness tasks, comparable to findings in young Chinese children (Arredondo et al. 2015). Similar activation patterns have also been reported in Chinese adults performing auditory morphological judgment tasks (Zou et al. 2015). Taken together, these studies highlight the left IFG as a core neural hub for morphological processing across languages and modalities. The present findings extend this view by demonstrating that even single-word processing in Chinese recruits left IFG activity when morphological constraints are involved, whether in auditory or visual contexts.

Beyond the left frontal regions, prior research has also implicated the left temporal cortex in morphological processing (Hsu et al. 2019; Ip et al. 2019). Although the present study did not reveal significant activation differences in the left temporal area across morphological conditions during Chinese word reading, subsequent correlation analyses uncovered reliable associations between temporal lobe activity and the late N400 component. This pattern suggests that the temporal cortex may contribute to the decoding of morphological structures, particularly when higher-order, controlled cognitive resources are recruited.

6.5 Conclusion

The present study combined behavioral, EEG, and fNIRS evidence to examine how morphological constraints are encoded during the reading of Chinese disyllabic compounds. Behaviorally, lexical decision results revealed clear effects of both semantic regularity and morphological legality, with pseudowords eliciting slower responses and lower accuracy compared to real words, reflecting the additional parsing cost of structurally irregular items. EEG analyses demonstrated that these effects unfold dynamically over time: early N400 effects indexed morphemic-level semantic analysis, whereas late N400 effects reflected the controlled evaluation of morphological legality. These findings further demonstrate that morphology and semantics can be dissociated in their neural representation, at least in terms of temporal dynamics, thereby deepening our understanding of the hierarchical structure within the mental lexicon. Complementary to this, fNIRS recordings localized these processes primarily to the left prefrontal cortex, corroborating the role of this region in managing both morphological and semantic constraints during Chinese word recognition.

Importantly, our findings lend further support to the dual-route account of Chinese word processing. Both whole-word and morphemic representations appear to be engaged, with their relative contributions depending on lexical properties such as structural legality. By integrating high-temporal-resolution EEG with spatially sensitive fNIRS, the study advances methodological approaches for investigating the neural mechanisms of morphology. Moreover, by situating Chinese results within broader cross-linguistic and cross-modal perspectives, the present work contributes to ongoing discussions about whether morphological processing in the brain is language-specific or governed by shared neurocognitive principles across writing systems and modalities.

Finally, the present study centered on Chinese compound words, which make up the majority of the Chinese lexicon. As a next step, Chap. 7 extends this investigation by contrasting compounds with derivations, thereby addressing the neural basis of different morphological types. Employing a priming paradigm alongside the same fused EEG–fNIRS approach, the following chapter aims to disentangle how structural differences between compounding and derivation are represented in the Chinese mental lexicon and instantiated in the brain.

References

Allen, M., Badecker, W., & Osterhout, L. (2003). Morphological analysis in sentence processing: An ERP study. *Language and Cognitive Processes*, *18*(4), 405-430.

Arredondo, M. M., Ip, K. I., Shih Ju Hsu, L., Tardif, T., & Kovelman, I. (2015). Brain bases of morphological processing in young children. *Hum Brain Mapp*, *36*(8), 2890-2900. https://doi.org/10.1002/hbm.22815

Baker, W. B., Parthasarathy, A. B., Busch, D. R., Mesquita, R. C., Greenberg, J. H., & Yodh, A. (2014). Modified Beer-Lambert law for blood flow. *Biomedical optics express*, *5*(11), 4053-4075.
Bick, A. S., Frost, R., & Goelman, G. (2010). Imaging implicit morphological processing: evidence from Hebrew. *Journal of Cognitive Neuroscience*, *22*(9), 1955-1969.
Bölte, J., Jansma, B. M., Zilverstand, A., & Zwitserlood, P. (2009). Derivational morphology approached with event-related potentials. *The Mental Lexicon*, *4*(3), 336-353.
Booth, J. R., Lu, D., Burman, D. D., Chou, T.-L., Jin, Z., Peng, D.-L., Zhang, L., Ding, G.-S., Deng, Y., & Liu, L. (2006). Specialization of phonological and semantic processing in Chinese word reading. *Brain research*, *1071*(1), 197-207.
Carrasco-Ortiz, H., & Frenck-Mestre, C. (2014). Phonological and orthographic cues enhance the processing of inflectional morphology. ERP evidence from L1 and L2 French. *Front Psychol*, *5*, 888. https://doi.org/10.3389/fpsyg.2014.00888
Chatrian, G., Lettich, E., & Nelson, P. (1985). Ten percent electrode system for topographic studies of spontaneous and evoked EEG activities. *American Journal of EEG technology*, *25*(2), 83-92.
Chee, M. W., O'Craven, K. M., Bergida, R., Rosen, B. R., & Savoy, R. L. (1999). Auditory and visual word processing studied with fMRI. *Human brain mapping*, *7*(1), 15-28.
Chung, K. K., Tong, X., Liu, P. D., McBride-Chang, C., & Meng, X. (2010). The processing of morphological structure information in Chinese coordinative compounds: an event-related potential study. *Brain Res*, *1352*, 157-166. https://doi.org/10.1016/j.brainres.2010.06.069
Feldman, L. B., Frost, R., & Pnini, T. (1995). Decomposing words into their constituent morphemes: Evidence from English and Hebrew. *Journal of Experimental Psychology: Learning, Memory, and Cognition*, *21*(4), 947.
Fu, G. Y., Mondloch, C. J., Ding, X. P., Short, L. A., Sun, L. P., & Lee, K. (2014). The neural correlates of the face attractiveness aftereffect: A functional near-infrared spectroscopy (fNIRS) study. *Neuroimage*, *85*, 363-371. https://doi.org/10.1016/j.neuroimage.2013.04.092
Greenhouse, S. W., & Geisser, S. (1959). On methods in the analysis of profile data. *Psychometrika*, *24*(2), 95-112.
Gu, J., & Yang, Y. (2010). A neuro-electrophysiological study on productivity of Chinese compounding [in Chinese]. *Yuyan Wenzi Yingyong* (3), 98-107.
Gu, J., Yu, L., & Ma, P. (2012). P2a reflects morphology structural processes in brain: Evidence from ERPs [in Chinese]. *Journal of Southwest University (Natural Science Edition)*, *34*(2), 137-144.
Hofmann, M. J., Dambacher, M., Jacobs, A. M., Kliegl, R., Radach, R., Kuchinke, L., Plichta, M. M., Fallgatter, A. J., & Herrmann, M. J. (2014). Occipital and orbitofrontal hemodynamics during naturally paced reading: An fNIRS study. *Neuroimage*, *94*, 193-202.
Hsu, C.-H., Pylkkänen, L., & Lee, C.-Y. (2019). Effects of morphological complexity in left temporal cortex: An MEG study of reading Chinese disyllabic words. *Journal of Neurolinguistics*, *49*, 168-177. https://doi.org/10.1016/j.jneuroling.2018.06.004
Hu, Z., Zhang, J., Couto, T. A., Xu, S., Luan, P., & Yuan, Z. (2018). Optical Mapping of Brain Activation and Connectivity in Occipitotemporal Cortex During Chinese Character Recognition. *Brain Topogr*, *31*(6), 1014-1028. https://doi.org/10.1007/s10548-018-0650-y
Huang, C. Y., Lee, C. Y., Huang, H. W., & Chou, C. J. (2011a). Number of sense effects of Chinese disyllabic compounds in the two hemispheres. *Brain Lang*, *119*(2), 99-109. https://doi.org/10.1016/j.bandl.2011.04.005
Huang, H. W., Lee, C. Y., Tsai, J. L., & Tzeng, O. J. (2011b). Sublexical ambiguity effect in reading Chinese disyllabic compounds. *Brain Lang*, *117*(2), 77-87. https://doi.org/10.1016/j.bandl.2011.01.003
Ip, K. I., Hsu, L. S., Arredondo, M. M., Tardif, T., & Kovelman, I. (2017). Brain bases of morphological processing in Chinese-English bilingual children. *Dev Sci*, *20*(5). https://doi.org/10.1111/desc.12449

Ip, K. I., Marks, R. A., Hsu, L. S., Desai, N., Kuan, J. L., Tardif, T., & Kovelman, L. (2019). Morphological processing in Chinese engages left temporal regions. *Brain Lang*, *199*, 104696. https://doi.org/10.1016/j.bandl.2019.104696

Jin, G., Xiao, H., & Fu, L. Z., Y.. (2005). The construction and deep processing of Modern Chinese Corpus [in Chinese]. *Applied Linguistics (China)* (2), 111-120.

Leminen, A., Smolka, E., Dunabeitia, J. A., & Pliatsikas, C. (2019). Morphological processing in the brain: The good (inflection), the bad (derivation) and the ugly (compounding). *Cortex*, *116*, 4-44. https://doi.org/10.1016/j.cortex.2018.08.016

Leong, C. K., Cheng, P. W., & Tan, L. H. (2005). The role of sensitivity to rhymes, phonemes and tones in reading English and Chinese pseudowords. *Reading and Writing*, *18*(1), 1-26.

Lin, X., Lei, V. L. C., Li, D., Hu, Z., Xiang, Y., & Yuan, Z. (2018). Mapping the small-world properties of brain networks in Chinese to English simultaneous interpreting by using functional near-infrared spectroscopy. *Journal of innovative optical health sciences*, *11*(03). https://doi.org/10.1142/s1793545818400011

Liu, L., Deng, X., Peng, D., Cao, F., Ding, G., Jin, Z., Zeng, Y., Li, K., Zhu, L., & Fan, N. (2009). Modality-and task-specific brain regions involved in Chinese lexical processing. *Journal of Cognitive Neuroscience*, *21*(8), 1473-1487.

Liu, L., Tao, R., Wang, W., You, W., Peng, D., & Booth, J. R. (2013). Chinese dyslexics show neural differences in morphological processing. *Dev Cogn Neurosci*, *6*, 40-50. https://doi.org/10.1016/j.dcn.2013.06.004

Luck, S. J. (2014). *An introduction to the event-related potential technique*. MIT press.

Newman, A. J., Ullman, M. T., Pancheva, R., Waligura, D. L., & Neville, H. J. (2007). An ERP study of regular and irregular English past tense inflection. *Neuroimage*, *34*(1), 435-445.

Noble, W. S. (2009). How does multiple testing correction work? *Nat Biotechnol*, *27*(12), 1135-1137. https://doi.org/10.1038/nbt1209-1135

Oldfield, R. C. (1971). The assessment and analysis of handedness: the Edinburgh inventory. *Neuropsychologia*, *9*(1), 97-113.

Pfeifer, M. D., Scholkmann, F., & Labruyere, R. (2017). Signal Processing in Functional Near-Infrared Spectroscopy (fNIRS): Methodological Differences Lead to Different Statistical Results. *Front Hum Neurosci*, *11*, 641. https://doi.org/10.3389/fnhum.2017.00641

Pinti, P., Tachtsidis, I., Hamilton, A., Hirsch, J., Aichelburg, C., Gilbert, S., & Burgess, P. W. (2020). The present and future use of functional near-infrared spectroscopy (fNIRS) for cognitive neuroscience. *Ann N Y Acad Sci*, *1464*(1), 5-29. https://doi.org/10.1111/nyas.13948

Pylkkänen, L., Feintuch, S., Hopkins, E., & Marantz, A. (2004). Neural correlates of the effects of morphological family frequency and family size: an MEG study. *Cognition*, *91*(3), B35-B45.

Pylkkänen, L., & Marantz, A. (2003). Tracking the time course of word recognition with MEG. *Trends in cognitive sciences*, *7*(5), 187-189.

Schremm, A., Novén, M., Horne, M., & Roll, M. (2019). Brain responses to morphologically complex verbs: An electrophysiological study of Swedish regular and irregular past tense forms. *Journal of Neurolinguistics*, *51*, 76-83. https://doi.org/10.1016/j.jneuroling.2019.01.006

Su, B. (2016). The implicity, diversity and cognitive features of Chinese compounding [in Chinese]. *Xueshu Yanjiu* (1), 162-165.

Wu, J., Chang, J., Qiu, Y., & Joseph, D. (2020). The temporal process of visual word recognition of Chinese compound: Behavioral and ERP evidences based on homographic morphemes [in Chinese]. *Acta Psychologica Sinica*, *52*(2), 113-127.

Xu, Y., Graber, H. L., Schmitz, C. H., & Barbour, R. L. (2014). *nirsLAB: A Problem-Solving Environment for fNIRS Neuroimaging Data Analysis*

Ye, J. C., Tak, S., Jang, K. E., Jung, J., & Jang, J. (2009). NIRS-SPM: statistical parametric mapping for near-infrared spectroscopy. *Neuroimage*, *44*(2), 428-447.

Zhang, B. (1993). Decomposed storage of Chinese disyllabic words in mental lexicon [in Chinese]. *Collection of Abstracts from the 9th Chinese Psychological Conference* Beijing.

Zhang, B., & Peng, D. (1992). Decomposed storage in the Chinese lexicon. In *Advances in Psychology* (Vol. 90, pp. 131-149). Elsevier.

Zhang, P., Zhang, Q., Peng, G., Song, X., & Bai, X. (2016). The Representation of Chinese Two-character Words in Mental Lexicon: Evidence from an fNIRS Study [in Chinese]. *Psychological Science*, *39*(4), 849-855.

Zhang, Y. (2019). *The research of abstract words semantic representation and brain mechanism for Tibetan-Chinese bilinguals* [in Chinese] (Thesis). Hebei Normal University.

Zhao, S., Wu, Y., Li, T., & Guo, Q. (2017). Morpho-semantic processing in Chinese word recognition: An ERP study. *Acta Psychologica Sinica*, *49*(3). https://doi.org/10.3724/sp.J.1041.2017.00296

Zou, L., Packard, J. L., Xia, Z., Liu, Y., & Shu, H. (2015). Neural Correlates of Morphological Processing: Evidence from Chinese. *Front Hum Neurosci*, *9*, 714. https://doi.org/10.3389/fnhum.2015.00714

Chapter 7
Neural Basis of Morphological Structure Effects in Chinese

7.1 Introduction

Chapters 5 and 6 examined the representational units of the Chinese mental lexicon and provided converging behavioral and neuroimaging evidence for the dual-route position, showing that Chinese morphological processing has clear neural reality and can be distinguished from semantic processing in the brain. Yet, both chapters focused exclusively on compounds, which constitute the majority of the Chinese lexicon. In those studies, compounding structures were proportionally matched to their distribution in natural corpora, rather than systematically manipulated. Building on this foundation, the present chapter aims to directly examine whether morphological structure effects can be observed in Chinese word processing.

From a broader linguistic perspective, morphology concerns not only how words are formed, but also how they are interconnected within the mental lexicon (Levelt 1989; Treisman 1960). Psycholinguistic and neurolinguistic research has provided accumulating evidence that sub-lexical information and morphological structure significantly shape lexical access. A long-standing debate over the past five decades has focused on whether morphologically complex words are stored and processed via full-listing or full-parsing mechanisms (De Grauwe et al. 2014; Frost et al. 2005; Jiang 2018; Schiller 2020; Wei et al. 2023). Regardless of whether decomposition is obligatory, it is well established that the way morphemes combine to form new words critically modulates word recognition. Against this backdrop, this chapter contrasts coordinate compounds and derivations to probe the cognitive and neural bases of morphological structure effects in Chinese. In doing so, it seeks to further evaluate the dual-route model and lexical processing theories such as spreading activation in the context of Chinese morphology.

F. Gao, *The When, Where, and How of Morphological Processing in the Human Brain*, https://doi.org/10.1007/978-981-95-5981-7_7

7.1.1 *Morphological Structure Effects in Alphabetical Languages*

To better understand the role of morphological structure, a large body of behavioral and neuroimaging research has investigated inflectional and derivational processing in alphabetic languages (Bölte et al. 2009; Carrasco-Ortiz and Frenck-Mestre 2014; Grainger and Beyersmann 2020; Leminen et al. 2019; Schiller 2020; Whiting et al. 2013). Inflections mark grammatical changes such as tense, number, or gender without altering the core meaning (e.g., "hikes" in "He hikes in mountains on weekends"), whereas derivations involve affixation that generates new words, meanings, or even categories (e.g., "hike-hiker"). ERP evidence has shown that irregular inflections can elicit anterior negativities (LANs), reflecting rule-based morphological computations, whereas P600 effects between regular and irregular violations suggested controlled processing of stored lexical forms (Newman et al. 2007). These findings can be interpreted within the framework of a dual route model (or "dual-access model") (Baayen et al. 1997; Bozic and Marslen-Wilson 2010; Clahsen 1999; Isel et al. 2003; Marslen-Wilson and Tyler 2007; Silva and Clahsen 2008), which emphasizes the distinct neurocognitive mechanisms underlying the processing of morphologically complex words. Within this dual framework, regular inflections in English are thought to primarily recruit the left fronto-temporal network, reflecting rule-based decomposition and recombination processes. In contrast, derivational forms and highly lexicalized words tend to engage a bilateral and more distributed network, supporting whole-word recognition and storage-based mappings between sound and meaning.

Compared to inflection and derivation, compounding has received less attention in alphabetic languages, even though it is a fundamental mechanism of morphological productivity across languages (Leminen et al. 2019; Libben 2014; Libben et al. 2020). Most psycholinguistic studies have examined how factors such as constituent frequency (e.g., Andrews et al. 2004), semantic transparency (e.g., Brooks and Cid de Garcia 2015; Frisson et al. 2008; Libben et al. 2003; Smolka and Libben 2017), and headedness (e.g., Libben et al. 2003; Marelli and Luzzatti 2012) influence compound word recognition, while little is known about the role of compound structure itself. Descriptive linguistic accounts (Jespersen 1914; Olsen 2012) distinguished a variety of structures, ranging from appositional (e.g., woman doctor) to argument-based (e.g., sunrise) and purpose-related (e.g., wineglass) ones. These classifications are partially based on syntactic relations yet mostly on semantics. Most psycholinguistic research on compound processing has concentrated almost exclusively on subordinate structures such as *blackbird* and *snowman* (Brooks and Cid de Garcia 2015; Smolka and Libben 2017). Although some studies have acknowledged the presence of coordinate or copulative compounds in English, typically expressed in hyphenated forms such as *singer-songwriter* and *architect-sculptor* (Chung et al. 2010; Olsen 2012), their neural underpinnings and potential contrasts with other compound subtypes remain unexplored. This lack of evidence likely reflects the relatively limited range of compounding patterns in English.

7.1.2 *Morphological Structure Effects in Chinese: Behavioral Evidence*

Unlike alphabetic languages, Chinese is typically regarded as a language with little inflection (though see Dai 1997 for debate), which sets it in sharp contrast to Indo-European languages. As summarized in Sect. 3.5 and drawing on structuralist linguistics (e.g., Huang and Liao 2011), Chinese vocabulary can be classified into three major structural types: monomorphemic words (e.g., 蝴蝶/hu2-die2/, "butterfly"), derivations (including prefixed and suffixed forms), and compounds, the latter constituting the largest proportion. Surveying modern Chinese dictionaries, Cao (2003) reported that in colloquial data, compounds comprise 83.02% of multimorphemic words, followed by derivations (14.51%). Within compounds, five major sub-structures have been identified: subordination (54%, e.g., 红茶/hong2 cha2/, "black tea"), coordination (26%, e.g., 山川/shan1 chuan1/, "mountains and rivers"), verb–object (18%, e.g., 吃饭/chi1 fan4/, "have a meal"), verb–resultative (2%, e.g., 提高/ti2 gao1/, "to boost"), and subject–predicate (1%, e.g., 地震/di4 zhen4/, "earthquake") (Su 2016), together with a special type of compounding reduplication that also contributes to Chinese morphological productivity.

One of the earliest psycholinguistic studies on Chinese morphological structure demonstrated that compound type modulates decomposition processes. B. Zhang and Peng (1992), for example, found that in coordinate compounds, the frequencies of both constituent morphemes influenced reaction times in word recognition, whereas in subordinate compounds, only the frequency of the final constituent exerted a significant effect. This suggested that morphological structure constrains the way compounds are accessed and parsed during reading. Following this seminal work, both behavioral and neuroimaging studies have provided converging evidence on the cognitive and neural mechanisms underlying morphemic effects in Chinese compound word processing. For instance, ambiguity and polysemy at the morphemic level have been shown to significantly affect the recognition of disyllabic compounds (Huang et al. 2011a,b; Tsang et al. 2013; Wu et al. 2020; Zou et al. 2019).

Beyond the role of individual morphemes, recent work has also examined inter-morpheme relations, revealing that both semantic/thematic connections (Ji and Gagné 2007; Jia et al. 2013; Pan and Jared 2020) and grammatical configurations shape how compound words are processed. These findings highlight that Chinese morphological structure effects are multifaceted, operating at the level of constituent morphemes as well as at the relational level between morphemes, and together they contribute to our understanding of how meaning is constructed in a morphosyllabic writing system.

In addition, prior research has primarily focused on subordinate and coordinate compounding structures, aiming to examine their representational differences in the mental lexicon. For example, Liu and McBride-Chang (2010b), using a compounding production task, compared the word production performance of third-grade children across four compound types. They found that subordinate and coordinate compounds were easier to generate than subject–predicate or verb–object compounds, and that this difficulty corresponded to the distributional characteristics of these structures in the real-world language (Laudanna and Burani 1995). Follow-up studies further distinguished subordinate and coordinate compounds through different paradigms. In a primed lexical decision task, subordinate structures were harder to recognize than coordinate ones when semantic relatedness and structural consistency between primes and targets were manipulated (Liu and McBride-Chang 2010a). Specifically, semantic priming was facilitated by structural congruency in subordinate compounds, whereas coordinate compounds showed the opposite trend. Moreover, subordinate structures have been shown to significantly enhance literacy performance in memorizing compound words compared with coordinate compounds (Liu 2016, 2017). Liu (2016) explained this advantage by appealing to the spreading activation theory (Collins and Loftus 1975), arguing that modifier–head relations within subordinate compounds provide stronger inter-constituent connections than those found in coordinate structures. As a result, coordinate compounds such as 花草 (/hua1-cao3/, flower-grass, "plant") demand more cognitive efforts because both constituents contribute equally but are only weakly connected. Building on this account, Liu (2017) further proposed the existence of an independent morphological structure layer in the mental lexicon, in addition to the established morpheme and word layers (Taft 1994). At this level, both semantic and syntactic connections between constituents are activated, but the weaker semantic associations in coordinate compounds may diminish their corresponding morphological effects.

Recent behavioral evidence further strengthens the view that morphological structure significantly shapes Chinese compound word recognition in both L1 and L2. Hao et al. (2024) conducted a series of experiments showing that morphological structure modulates both the speed and accuracy of processing, highlighting how L2 learners' morphological awareness contributes to lexical representation. Importantly, subordinate and coordinate compounds elicited distinct processing patterns, with subordinate compounds generally being more facilitative in recognition than coordinate ones, consistent with the spreading activation account of stronger modifier–head connections.

Sun, Pae, and colleagues extended this line of inquiry by systematically comparing the recognition of coordinate compounds among Chinese native speakers and L2 learners (Sun et al. 2020, 2021). More recently, Sun et al. (2024) broadened the scope by investigating all five major compound structures (subordinate, coordinate, verb–object, verb–complement, and subject–predicate) among English- and Korean-speaking learners of L2 Chinese, compared with native Chinese readers. Their findings revealed main effects of both L1 background and morphological structure. Specifically, the subject–predicate structure consistently emerged as the most difficult for both L1 and L2 readers, likely due to its low frequency and lack of structural equivalence in English, and only partial overlap in Korean. By contrast, verb–object and subordinate compounds were processed more efficiently. Notably, Korean learners responded faster than English learners across all structures, though at the cost of reduced accuracy, suggesting that orthographic and morphological similarities between Korean and Chinese afforded an advantage in initial recognition but not necessarily in precise lexical judgment.

Together, these studies converge on the idea that Chinese morphological structures exert systematic effects on lexical processing across native and non-native readers, and that the relative difficulty of different structures is constrained by both distributional properties within Chinese and cross-linguistic transfer from learners' L1. This accumulating evidence points to the need for a refined neurocognitive model of Chinese morphology that accounts for structural diversity, distributional frequency, and cross-linguistic variation.

7.1.3 Morphological Structure Effects in Chinese: Neural Evidence

While behavioral studies (e.g., Liu and McBride-Chang 2010a) reported distinct patterns between subordinate and coordinate structures, these differences were not consistently replicated in ERP studies. For instance, Chung et al. (2010) and Cui et al. (2018) failed to detect parallel behavioral contrasts in ERP paradigms. One possible explanation is that Liu et al.'s work adopted a relatively longer SOA of 200 ms, which may have shifted the priming effect from facilitation to inhibition, such that coordinate compounds showed inhibitory rather than facilitatory priming when prime and target shared the same structure. By contrast, Chung et al.'s ERP study, which also examined coordinate compounds, reported a facilitatory effect when primes and targets shared the same structure. Specifically, structural priming pairs elicited greater P250 (220–300 ms) amplitudes than structurally distinct pairs, suggesting a top-down facilitation of word structure during the early stage of lexical access (Hill et al. 2002). This discrepancy highlights the need for further investigation into whether and how morphological structure modulates the recognition of Chinese coordinate compounds.

Neuroimaging evidence on this issue is still scarce. Only a limited number of studies have examined the neural correlates of Chinese morphological processing in general (Ip et al. 2017, 2019; Zou et al. 2015). These studies consistently implicated both the left frontal and temporal cortices but did not distinguish between compound sub-structures and derivations. To date, only one functional neuroimaging study has explicitly addressed differences among compound subtypes. Using a lexical decision task and MEG, Hsu et al. (2019) contrasted mono-morphemic and multi-morphemic words including subordinate, coordinate, and verb–object compounds. Their findings showed that compound words elicited stronger activations in the left temporal cortex around 200 ms. Crucially, during the later 300–400 ms window, subordinate and verb–object compounds produced greater activation in the posterior temporal region than mono-morphemic words, whereas coordinate compounds did not differ significantly from the baseline. The authors attributed the null effect for coordinate structures to their lack of clear specifier–head–complement relations, as both constituent morphemes contribute equally and loosely to overall meaning.

Nevertheless, this work did not include the frontal cortex as a region of interest, even though prior studies identified the left IFG as a central hub in Chinese morphological processing (Ip et al. 2017; Zou et al. 2015). Thus, it remains an open question whether frontal regions, in concert with temporal cortices, also play a role in differentiating compound subtypes, underscoring the need for multimodal neuroimaging evidence.

Despite accumulating evidence for morphological processing in Chinese, the neurobiological underpinnings of different morphological structures remain poorly understood, both in terms of their temporal signatures and cortical localizations. In particular, the neural reality of coordinate compounds is still debated, as previous studies have yielded somewhat inconsistent findings regarding their processing advantage or disadvantage. Moreover, a striking gap in the literature is that no prior neuroimaging study has systematically examined the brain representation of Chinese derivational words, even though derivations constitute the second most frequent morphological type in Chinese vocabulary after compounds.

The present study is therefore designed to address these gaps by exploring whether Chinese derivations recruit cognitive and neural resources similar to those of compound words, and by examining the extent to which morphological priming effects can be distinguished from purely semantic (non-morphological) associations. Specifically, we focus on root priming, defined as the processing advantage that arises when prime and target share the same morpheme, in both derivational and compound words, contrasted with non-morphological semantic priming. To ensure a clear structural contrast, only coordinate compounds are selected for comparison with derivations. This allows us to test whether the brain encodes morphological structure beyond general semantic relatedness.

To this end, we adopt a masked priming paradigm in a visual lexical decision task, which is well-suited to tap into early, automatic morphological parsing (Fiorentino et al. 2014). Neural activity is recorded simultaneously with EEG, to capture millisecond-level temporal dynamics, and fNIRS, to map spatial patterns of cortical activation. The combined use of these two complementary techniques provides a more comprehensive picture of the spatio-temporal brain mechanisms underlying Chinese derivational and compound word processing.

7.2 Methods

7.2.1 Participants

Thirty native speakers of Mandarin Chinese (15 females; mean age = 22.2 years, SD = 3.2) participated in the study. All were undergraduate or graduate students from diverse majors at the University of Macau, and none majored in linguistics.

Participants were right-handed, had normal or corrected-to-normal vision, and reported no history of neurological or psychiatric disorders. The study protocol was approved by the Institutional Review Board of the University of Macau, and written informed consent was obtained from all participants prior to testing.

7.2.2 *Materials*

Thirty-six disyllabic Chinese compound words with coordinate structures (e.g., 江湖/jiang1 hu2/, "rivers and lakes") and another 36 disyllabic derivational words containing common suffixes (e.g., -家/jia1/, "-expert") were selected as target stimuli (see the full lists in Appendix D). In a masked priming paradigm (Lavric et al. 2007), each of these words was preceded by a prime consisting of its head morpheme. Additionally, another set of 36 disyllabic compounds involving four lexical structures (subordinate: 22, coordinate: 8, verb–object: 5, noun–complement: 1) served as a control group, where primes were single-character words semantically related to the targets. For this control group, the prime and target shared no morphological relationship and differed at both orthographic and phonological levels. To validate semantic consistency, 12 native Chinese speakers rated the semantic relatedness of the prime–target pairs across the derivation, compound, and control sets on a 7-point Likert scale. Mean ratings in all three groups exceeded five and showed non-significant difference ($p > 0.05$), confirming high semantic transparency and comparability. Moreover, the three sets of words were matched on lexical frequency, character frequency, and stroke number (Table 7.1). Finally, 108 disyllabic Chinese nonwords were selected from the MELD-SCH database dataset (Y. K. Tsang et al. 2018). These pronounceable but meaningless items were primed by unrelated characters and served as fillers to balance yes/no responses. All experimental stimuli were provided in Appendix D.

Table 7.1 Lexical characteristics and examples across the three conditions

	Prime	Frequency	Stroke Number	Target	Frequency	Stroke Number	Semantic Transparency
Derivational word	画/hua4/, paint	94 (145)	8.3 (2.3)	画家/hua4 jia1/, painter	8 (32)	15.3 (3.8)	5.2 (0.9)
Compound word	江/jiang1/, river	83 (109)	8.6 (3.1)	江湖/jiang1 hu2/, rivers and lakes	7 (10)	16.9 (4.6)	5.4 (0.9)
Non-morphological	饭/fan4/, meal	356 (1076)	8.8 (2.8)	早餐/zao3 can1/, breakfast	34 (79)	17.8 (4.5)	5.3 (0.5)

7.2.3 Procedure

A visual lexical decision task was employed in this study, adapted from the morphological priming paradigm with a masked priming procedure and a short SOA (Chung et al. 2010; Lavric et al. 2007). The experiment was programmed in E-prime (Fig. 7.1). Each trial began with a white fixation cross presented at the center of the black screen for 300 ms, followed by a 200 ms blank. Next, a string of asterisks was shown for 300 ms as a foreword mask, which was then replaced by the prime for 57 ms. Immediately afterward, the target word appeared, and participants were required to judge as quickly as possible whether the character string was a legitimate Chinese word by pressing the designated "yes" or "no" keys. Responses exceeding three seconds were automatically recorded as incorrect. A blank interval of 1–4 seconds followed each trial. Unlike the slow event-related design adopted in Chap. 6, here we employed a rapid event-related design in the fNIRS recordings (Schaeffer et al. 2014; Zhang et al. 2016), which allows for denser trial presentation and captures hemodynamic responses with higher temporal efficiency. Real-word trials comprised three lexical conditions. For display, primes were shown in Italic Kaiti font (size 40), whereas targets were presented in bold SimHei (size 40). Before the formal session, participants completed ten practice trials with accuracy feedback.

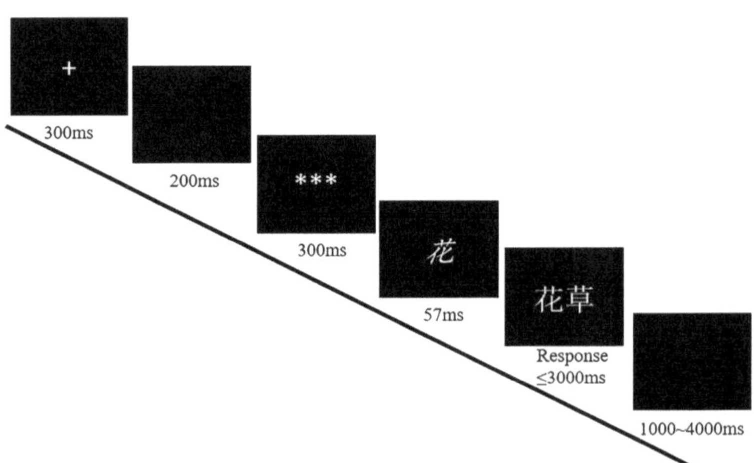

Fig. 7.1 Schematic demonstration of the lexical decision task with morphological priming technique and a short SOA

7.2.4 EEG-fNIRS Data Acquisition and Analysis

Similar to Chap. 6, both EEG and fNIRS signals were recorded simultaneously using an EasyCap system (Brain Products, Munich, Germany), linked to both the Brain Products EEG amplifier and the NIRScout device (NIRx Medizintechnik GmbH, Berlin, Germany). For EEG acquisition, 32 electrodes were mounted according to the international 10–20 system, with signals sampled at 500 Hz and filtered online between 0.03–70 Hz. The left mastoid served as the online reference, and electrode impedances were maintained below 20 kΩ. Offline preprocessing was conducted in EEGLAB v2021.0. Data were re-referenced to the average of all channels and band-pass filtered from 1–30 Hz. Noisy channels, accounting for less than 3% of the total, were reconstructed using spherical spline interpolation. Epochs were extracted from −157 ms to 1000 ms relative to target onset. The baseline was chosen to align with the short SOA in the masked priming design, ensuring that the baseline period captured brain activity prior to prime presentation rather than overlapping with priming effects (Chung et al. 2010). Ocular artifacts were identified and removed through ICA method with the ICLabel plugin (Pion-Tonachini et al. 2019), followed by manual rejection of residual contaminated epochs. Mean amplitudes were then extracted from predefined electrode clusters within two critical time windows, 220–300 ms (P250) and 300–500 ms (N400), based on previously reported ERP components (Chung et al. 2010). For the P250 analysis, electrodes AFF5h, FC1, FCz, FC2, and FC6 over the frontal midline and bilateral regions were selected, whereas the N400 was assessed using centro-parietal sites CPP5h, CP1, CP2, and CP6 (see Fig. 7.2).

For fNIRS data acquisition, eight LED sources and eight detectors were positioned over the left frontal and temporal cortices, producing a total of 22 measurement channels (Fig. 7.2). The optodes emitted light at 760 nm and 850 nm, with an inter-optode spacing of 3 cm. Optical signals were sampled at 7.81 Hz. Spatial registration was conducted by mapping the optodes to the international 10–20 system, after which the MNI coordinates of all sources, detectors, and channels were derived and imported into the NIRS-SPM toolbox (Ye et al. 2009) to obtain anatomical labels and overlap indices. After data obtained, preprocessing was carried out in nirsLAB (Xu et al. 2014). Data from one participant were excluded due to excessive physiological noise, leaving 29 valid datasets. Each trial was modeled for a 14 s period, including a 1 s baseline before target onset and a 13 s post-stimulus window to capture the hemodynamic response. Motion artifacts were corrected using nirsLAB's built-in procedures, and signals were band-pass filtered between 0.01–0.2 Hz to reduce drift and respiratory noise. Changes in HbO and HbR were estimated using the GLM within the SPM Level 1 module, applying a canonical HRF. The resulting beta coefficients were extracted for each priming condition at the individual level.

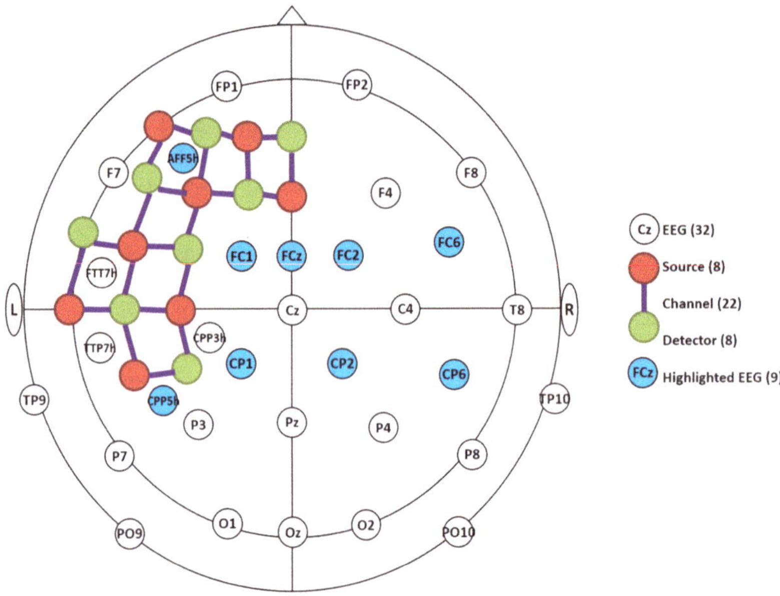

Fig. 7.2 The illustration of fused EEG-fNIRS arrangement and the highlighted EEG sites for ERP analysis

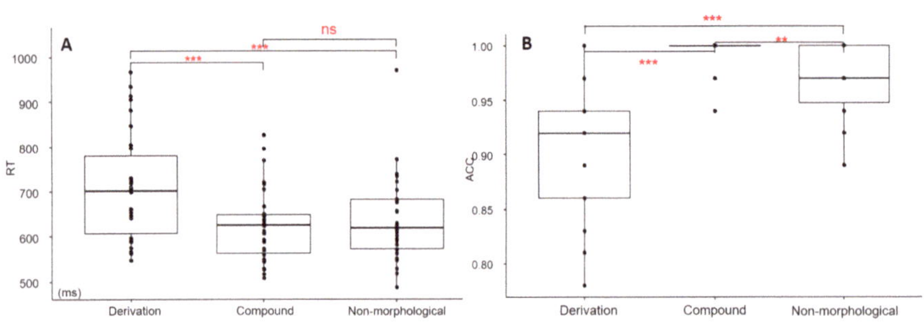

Fig. 7.3 RT (ms) and ACC for each of the three priming conditions

7.3 Results

7.3.1 *Behavioral Results*

The overall accuracy rate reached 95.26%, suggesting that participants were highly engaged with the task. Repeated-measures ANOVAs were then performed on both RT and ACC. For RT, a significant priming effect emerged, F (2,58) = 34.085, $p < 0.001$, partial $\eta^2 = 0.540$. Post hoc comparisons (Fig. 7.3a) showed that responses under derivational priming (712 ± 121 ms) were significantly slower than under compound priming (625 ± 82 ms) and non-morphological priming (636 ± 99 ms) (all $ps < 0.001$), whereas no reliable difference was found between compound and non-morphological conditions ($p > 0.05$).

A parallel effect was observed for ACC, F (2,58) = 37.873, $p < 0.01$, partial $\eta^2 = 0.566$. Accuracy in the derivational priming condition (89.87 ± 6.19%) was significantly lower than in the compound (98.8 ± 2.31%) and non-morphological (97.1 ± 3.03%) conditions (all $ps < 0.001$). In addition, non-morphological priming yielded a modest but significant reduction in ACC compared with compound priming ($p < 0.01$). The behavioral outcomes are summarized in Fig. 7.3. The significance of multiple comparisons was provided (**$p < 0.01$, ***$p < 0.001$).

7.3.2 ERP Results

The P250 component was analyzed in the 220–300 ms window following target word onset across the frontal bilateral and midline electrodes. A two-way repeated-measures ANOVA was conducted with priming type and hemisphere as within-subject factors at the bilateral sites (left: AFF5h, FC1; right: FC2, FC6). Although no overall effect of priming condition reached significance, F (2,58) = 2.169, $p = 0.131$, partial $\eta^2 = 0.07$, a robust main effect of hemisphere was observed, F (1,29) = 0.252, $p < 0.05$, partial $\eta^2 = 0.153$, together with a significant interaction between priming type and hemisphere interaction. Specifically, the right hemisphere (2.55 ± 0.31 μV) elicited larger P250 amplitudes than the left (1.92 ± 0.30 μV). In the left hemisphere, the non-morphological priming condition (2.10 ± 0.33 μV) produced greater P250 than the compound condition (1.54 ± 0.37 μV), whereas in the right hemisphere, derivational priming (2.97 ± 0.37 μV) evoked significantly higher P250 than the compound case (2.27 ± 0.34 μV). No other comparisons reached significance. At the midline site (FCz), no reliable differences among priming conditions were found, F (2,58) = 1.262, $p = 0.289$, partial $\eta^2 = 0.042$.

For the N400 (300–500 ms), guided by previous work and grand-averaged waveforms and topographies (Figs. 7.4 and 7.5), analyses were conducted with priming type and hemisphere (left: CPP5h, CP1; right: CP2, CP6) as factors. Results revealed significantly reduced N400 amplitudes in the left hemisphere (0.295 ± 0.268 μV) relative to the right (1.756 ± 0.278 μV), F (1,29) = 40.433, $p < 0.001$, partial $\eta^2 = 0.582$. No additional main effects or interactions emerged.

To further clarify the differences between morphological structures, priming effects were indexed by subtracting ERP amplitudes of control trials from those of compound and derivational trials (compound–control vs. derivation–control). Derivational priming (0.279 ± 0.350 μV) yielded significantly greater P250 amplitudes than compound priming (−0.354 ± 0.250 μV) across hemispheres ($ps < 0.05$). Moreover, right-hemisphere P250 (0.206 ± 0.294 μV) was significantly larger than the left (−0.280 ± 0.261 μV) ($p < 0.05$), a pattern consistent with midline electrodes. For the N400, no clear distinctions were observed across the three priming conditions.

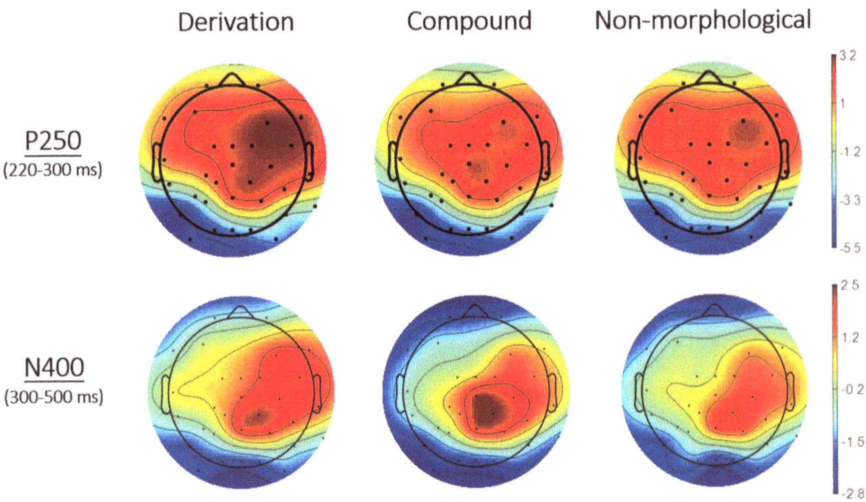

Fig. 7.4 Brain topographies of P250 and N400 effects across three conditions

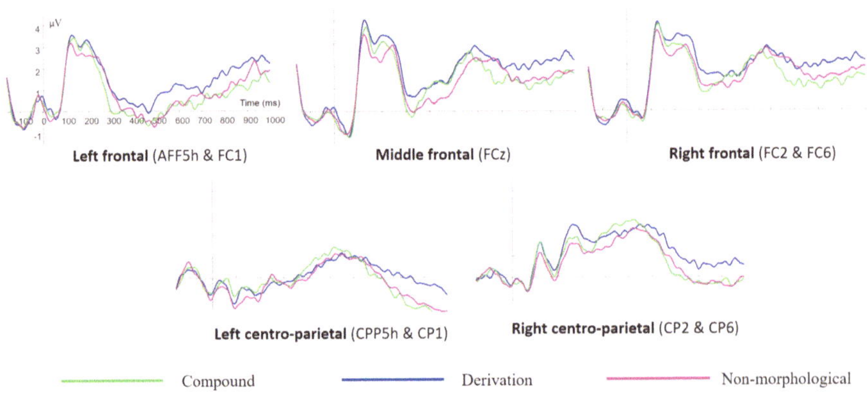

Fig. 7.5 Grand averaged ERPs over the frontal and centro-parietal sites

7.3.3 fNIRS Results

Similar to Chap. 6, the present analyses focused solely on HbO signals, as they are considered a more sensitive index of task-related neural activity. A series of repeated-measures ANOVAs were conducted channel by channel across the three priming conditions. Among these, a significant priming type effect was identified in channel 6, located in the frontopolar region, $F(2,56) = 3.162$, $p < 0.05$, partial $\eta^2 = 0.101$. Post-hoc pairwise comparisons were then performed for Derivation vs. Compound, Derivation vs. Non-morphological, and Compound vs. Non-morphological comparisons. The results are summarized in Table 7.2, which reports both significant ($p < 0.05$) and marginal effects ($0.05 < p < 0.1$, uncorrected).

Table 7.2 Comparison results of significant and marginally significant channels

CH#	MNI coordinates			BA	Anatomical label	Overlap	Comparisons	*t*	*p* (uncorrected)	*p* (FDR corrected)
	x	y	z							
5	−21	67	53	9	Dorsolateral prefrontal cortex	0.8178	Compound > Non-morphological	2.42	0.02	0.07
6	−44	67	27	10	Frontopolar area	0.95455	Compound > Non-morphological	2.84	0.01	0.02
8	−41	58	47	9	Dorsolateral prefrontal cortex	0.88688	Compound > Non-morphological	1.93	0.06	0.19
10	0	73	49	9	Dorsolateral prefrontal cortex	0.75591	Compound > Derivation	1.99	0.06	0.16
11	−15	61	61	8	Includes Frontal eye fields	0.77686	Compound > Non-morphological	2.07	0.05	0.14
18	−67	6	58	6	Pre-Motor and Supplementary Motor Cortex	0.92913	Compound > Non-morphological	1.74	0.09	0.28

Specifically, enhanced activation was observed in the left frontal cortex, including pre-motor and supplementary motor areas (channels 5, 6, 8, 11, and 18), when comparing non-morphological priming with compounds. In addition, compounds elicited stronger activation than derivations in channel 10. These activation patterns are illustrated in Fig. 7.6.

To further probe the word structure effect, differential contrasts (derivation/compound minus non-morphological) were examined with t-tests. Notably, only channel 10 approached significance ($p = 0.057$), showing that compounds tended to induce greater activation than derivations. This suggests that compound words may recruit stronger frontal resources than derivational words during morphological processing, highlighting possible differences in cognitive demand between the two structures.

7.3.4 *Correlational Results*

Pearson correlation analyses were carried out to examine the relationships between ERP P250 amplitudes and fNIRS HbO beta values. For frontal activity, the mean amplitudes from EEG electrodes FP1, F7, AFF5h, FC1, and FCz were compared with hemodynamic responses from fNIRS channels 5, 6, 8, 10, and 11. Temporal activity was assessed by pairing EEG sites (FTT7h, TTP7h, CPP5h, CPP3h, CP1) with fNIRS channels 15, 16, 17, 21, and 22. The resulting correlation matrices (see Fig. 7.7) showed that P250 amplitudes were strongly predictive of temporal cortical activation during derivational word reading, pointing to the engagement of an extended semantic network. Notably, significant correlations between P250 and HbO responses were also observed in both frontal and temporal regions, particularly in the derivation–compound contrast, suggesting shared neural dynamics across electrophysiological and hemodynamic measures. In Fig. 7.7, regions with brighter colors indicate stronger correlations, while the dashed black square highlights the region of interest. * denotes $p < 0.05$.

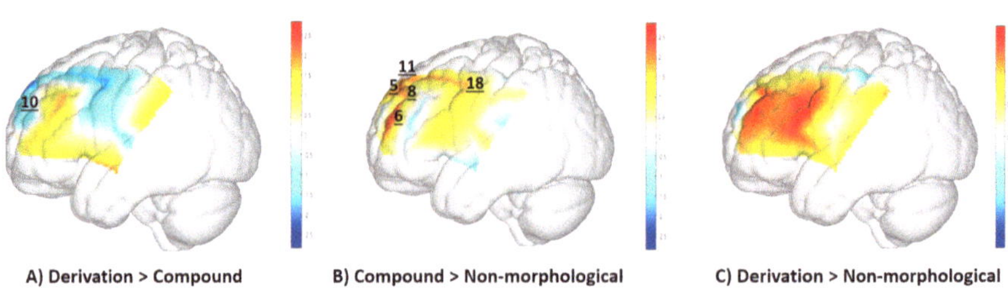

Fig. 7.6 T maps and significant channels of three pairwise comparisons as revealed by fNIRS data

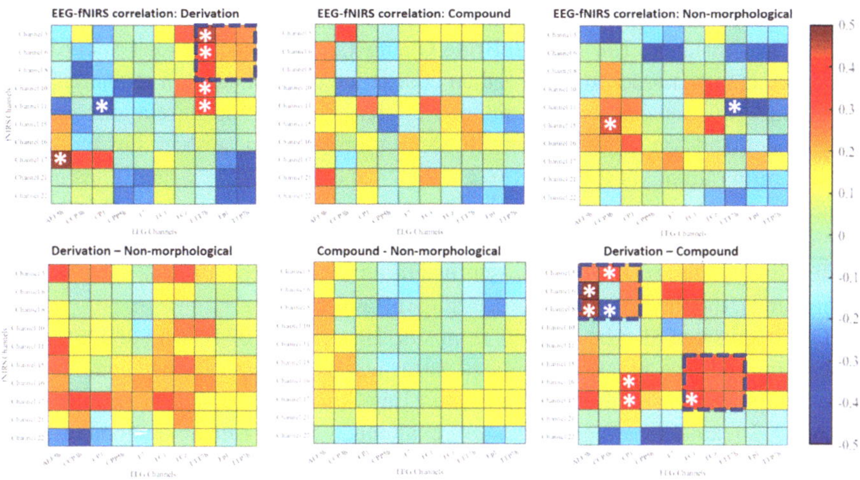

Fig. 7.7 Matrices of correlation coefficients linking ERP responses with fNIRS-based brain activation

7.4 Discussion

The present experiment investigated both morphological priming (compound or derivational root priming vs. non-morphological semantic priming) and word structure effects (derivation vs. compound) using a masked priming lexical decision paradigm. The behavioral analyses showed clear facilitation for coordinate compounds: accuracy was significantly higher and reaction times shorter than in the non-morphological condition. This result extends Chung et al. (2010) by demonstrating that priming benefits in coordinate words can emerge not only from pairs sharing the same overall structure, but also when the first morpheme serves as a prime, provided that the SOA is very rapid (57 ms). However, these findings contrast with those of Liu and McBride-Chang (2010a) and Cui et al. (2018), where longer SOAs (200 ms) or different task requirements (e.g., self-paced judgments) produced inhibitory or null effects. Together, the current evidence suggests that the coordinate structure effect in Chinese is highly sensitive to SOA and may reflect an automatic, short-lived activation of morphological information at the earliest stage of lexical access. In contrast, no reliable facilitation from morphological priming was observed for Chinese derivations. Since derivational morphology in Chinese has rarely been tested for its psychological validity, and because structural effects are often attenuated in behavioral tasks due to methodological constraints (Chung et al. 2010; Cui et al. 2018), stronger evidence may need to come from electrophysiological and neuroimaging data, which can better capture the neural correlates of such subtle processes.

With respect to morphological structure effects, both RT and accuracy data suggested that derivations were more difficult to recognize than coordinate compounds in constituent priming conditions. This finding echoes earlier evidence from compound processing, where structural relations were shown to influence lexical access. For instance, Liu and McBride-Chang (2010a) reported that subordinate compounds were harder to process than coordinate compounds in a semantic priming paradigm, consistent with the notion that modifier–head relations impose heavier demands on structural parsing. Derivations in Chinese share this modifier–head configuration, yet the suffix (e.g., -家 "-expert" in 作家 "writer") has undergone grammaticalization, functioning as a bound morpheme with weaker lexical connectivity than the two constituents of a coordinate compound. As a result, base–suffix connections may be even looser than those between constituents in coordinate compounds, thereby increasing processing difficulty.

Our results can also be situated alongside Hao et al. (2024), who demonstrated that Chinese native speakers exhibited faster responses to coordinate and subordinate compounds than other structures, and Jing Sun et al. (2024), who showed that L1 and L2 learners were generally less challenged by coordinate compounds than subject-predicate and verb-object. Taken together, these studies highlight that morphological structure significantly modulates recognition difficulty, with coordinate compounds generally easier to process than some other compounding ones, and derivations posing even greater challenges due to their weak morphemic connections. This broader pattern supports the spreading activation account, in which the strength of morpheme-to-morpheme associations governs cognitive effort in word recognition.

Given the behavioral evidence, the ERP findings provided converging support for structural differences in Chinese word processing. Specifically, analyses of the frontal P250 revealed a robust morphological structure effect: derivations elicited significantly larger positivities than coordinate compounds, with the effect being more pronounced in the right hemisphere. This frontal component has previously been linked to sensitivity to semantic relatedness and to structural consistency between primes and targets (Chung et al. 2010; Hill et al. 2002). Importantly, prior research has shown that the P250 emerges only under short SOA conditions (e.g., 57 ms or 150 ms) but disappears when the SOA is prolonged to 700 ms, suggesting that it indexes automatic access during lexical processing. Consistently, MEG studies (e.g., Cavalli et al. 2016) have associated this early time window with rapid decomposition and parsing operations in morphological analysis.

In the present study, the stronger P250 response for derivations compared to coordinate compounds implies that derivational words require greater cognitive resources during early morphological parsing. This pattern mirrors the behavioral results, highlighting that structural differences at the morphemic level directly shape processing demands. Within the framework of spreading activation, word roots can be understood as nodes connected with varying associative strengths. Upon presentation of a prime, related morphemes and potential recombinations are simultaneously activated, after which structural re-composition occurs. Yet the associative strength differs across structures: while coordinate compounds involve two morphemes contributing equally to the whole, their interconnection is relatively loose compared to modifier–head subordinations (Liu 2016), yet stronger than base–suffix pairs in derivations. The current P250 effects therefore suggest that the frontal cortex is sensitive enough to distinguish derivations from compounds, reflecting their distinct morphological organization within the mental lexicon.

Beyond the P250 effect, Chung et al. (2010) also identified a canonical N400 modulation driven by semantic relatedness between primes and targets. By contrast, the present study did not replicate such a semantic priming effect. Although the grand-average waveforms and scalp topographies suggested an N400-like negativity broadly distributed over the centro-parietal region, statistical analyses revealed no significant differences across the three priming conditions. A plausible explanation is that semantic associations were not systematically manipulated in our design, since primes and targets across conditions were matched in semantic relatedness. Consequently, the frontal P250 observed here may reflect a relatively "cleaner" effect of morphological structure, disentangled from semantic influences.

Importantly, when considered together with the findings of Chap. 6, the present results further support the relative dissociation between semantic and morphological processing in Chinese word recognition. Unlike Chap. 6, where semantic effects emerged earlier than morphological encoding, the current study revealed that morphological processing (P250) preceded semantic integration (N400). We therefore speculate that the frontal P250 may reflect a task-specific feature of the priming paradigm, indexing early attentional allocation and structural pre-activation (Chung et al. 2010; Gu et al. 2012; Hill et al. 2002), whereas semantic processing arises at a later stage. In line with the dual-route model of morphological processing, this early P250 effect might correspond to a decomposition-based route that operates automatically at sub-lexical levels, while later semantic components reflect whole-word access and integration.

The fNIRS findings offered converging neural evidence for the morphological priming effect and word structure effect in compound word reading. Specifically, coordinate compounds elicited stronger hemodynamic responses than derivation and control conditions in the left frontal network, particularly in the DLPFC and frontopolar regions. This frontal activation pattern aligns broadly with earlier fMRI research on morphological processing in children and auditory tasks. For example, Zou et al. (2015) reported that morpheme-sharing word pairs (e.g., 大风/da4 feng1/ – 大门/da4 men2/, "big wind–big door") produced the strongest activity in the left IFG during an explicit auditory morphological judgment task. Such results underscore the left IFG as a critical hub for morphological processing not only in Chinese but also across alphabetic languages (Bozic et al. 2007; Lehtonen et al. 2006).

Similarly, Ip et al. (2017) demonstrated that Chinese–English bilingual children recruited the left IFG more strongly in an auditory English derivational task (e.g., jump-rejump) than in a Chinese compound morphology task (e.g., 病人/bing4 ren2/ - 病花/bing4 hua1/, "sick man-sick flower"), a pattern resembling that of English monolinguals. These converging findings further highlight the functional role of the left frontal cortex in supporting morphological operations across languages and literacy stages. Although fNIRS has limited spatial precision compared with fMRI, which may explain the lack of precise localization in the IFG in the present study, the robust activations observed in the DLPFC nevertheless reinforce the view that the frontal cortex constitutes a central node for morphological processing in print reading among adult readers.

Although the present study did not cover the full range of Chinese word structures, it nonetheless provided preliminary evidence for a structural contrast between derivations and coordinate compounds, thereby making an original contribution to the field. The seemingly divergent findings between ERP and fNIRS results may actually reflect complementary aspects of morphological processing at different stages and levels of analysis. The ERP data revealed that derivations elicited larger P250 amplitudes than compounds, suggesting that derivational words require more efforts in the early parsing of morpheme–affix combinations. This is likely due to the relatively weaker semantic binding between base and affix, as affixes in Chinese (e.g., –家, –儿) often undergo grammaticalization and contribute less transparent meaning, thereby demanding additional cognitive resources during early structural analysis. In contrast, the fNIRS results showed stronger left prefrontal activation for compounds than for derivations. Given that the DLPFC is closely associated with higher-level cognitive control, monitoring, and integration (e.g., Butler et al. 2023; Mitchell et al. 2016), this pattern indicates that compound words, especially coordinate structures in which both morphemes contribute equally to the overall meaning, engage more sustained integration and executive processing at later stages. Taken together, these findings suggest a division of labor: derivations impose greater demands on early automatic parsing processes (as reflected by P250), whereas coordinate compounds place heavier burdens on later frontal integration mechanisms (as captured by hemodynamic responses in fNIRS).

In addition, correlational analyses comparing derivations and compounds revealed a coherent pattern linking frontal P250 amplitudes with hemodynamic activity in the left frontal regions. Collectively, these findings converge with those reported in Chap. 6, where the left frontal cortex was implicated in sensitivity to morphological constraints. While Chap. 6 demonstrated that the frontal cortex automatically detects the legality of compound structures (constraint encoding), the current results extend this role by showing that the same region is also engaged in distinguishing different morphological structures (structure effect). This comparison suggests that the left frontal cortex constitutes a core hub for morphological processing in Chinese reading, supporting both the detection of structural constraints and the differentiation of morphological types across tasks and paradigms.

Moreover, current neuroimaging findings also lend support to the dual-route theory of morphological processing in morphologically rich languages (Baayen et al. 1997; Bozic and Marslen-Wilson 2010). This account proposes that the recognition of regular inflections and rule-based word units relies on an online combinatorial mechanism, primarily engaging the left fronto-temporal network. In contrast, highly lexicalized and storage-based items are processed via broader, often bilateral, brain regions, which enables more efficient access with reduced computational demands. Within Chinese, compounding represents the dominant morphological pattern, and its prevalence suggests a relatively storage-oriented representation in the mental lexicon. By comparison, although derivations constitute a smaller proportion of the lexicon, they are nevertheless highly productive due to the adhesive nature of affixes. The observed neural differences between compounding and derivation may thus be explained by the varying strength of associative links between morphemic constituents, consistent with the predictions of the spreading activation theory.

7.5 Conclusion

In sum, this study investigated both the morphological priming effect and the structural contrast between derivations and coordinate compounds by employing simultaneous EEG–fNIRS recordings. The results revealed a clear word structure effect in the left frontal cortex. Specifically, derivational words imposed heavier demands on early, automatic parsing processes, as reflected by enhanced P250 responses, whereas compounds elicited stronger activation in later frontal integration mechanisms, as indexed by hemodynamic changes in fNIRS. These findings extend the evidence from Chap. 6 by highlighting the central role of the left frontal cortex in Chinese morphological processing and further support the applicability of the dual-route model and the spreading activation framework to Chinese.

A limitation of the present work lies in the restricted range of morphological structures tested, as compared to prior behavioral studies that included a broader set of subtypes. Nonetheless, the converging ERP and fNIRS evidence contributes novel insights into the neurocognitive underpinnings of morphological structure effects in Chinese word reading. Importantly, the contrast between derivations and compounds also underscores how the morpho-syllabic nature of Chinese shapes the balance between storage-based and computation-based lexical processing, enriching cross-linguistic models of morphological representation.

Finally, building on these findings, Chap. 8 will turn to the bilingual context to examine whether the neural resources underlying Chinese morphological processing can be transferred across languages in learning settings, thereby offering a window into the cross-linguistic neurocognitive mechanisms of morphology.

References

Andrews, S., Miller, B., & Rayner, K. (2004). Eye movements and morphological segmentation of compound words: There is a mouse in mousetrap. *European Journal of Cognitive Psychology, 16*(1-2), 285-311.

Baayen, R. H., Dijkstra, T., & Schreuder, R. (1997). Singulars and plurals in Dutch: Evidence for a parallel dual-route model. *Journal of Memory and Language, 37*(1), 94-117.

Bölte, J., Jansma, B. M., Zilverstand, A., & Zwitserlood, P. (2009). Derivational morphology approached with event-related potentials. *The Mental Lexicon, 4*(3), 336-353.

Bozic, M., Marslen-Wilson, W. D., Stamatakis, E. A., Davis, M. H., & Tyler, L. K. (2007). Differentiating morphology, form, and meaning: Neural correlates of morphological complexity. *Journal of Cognitive Neuroscience, 19*(9), 1464-1475.

Bozic, M., & Marslen-Wilson, W. (2010). Neurocognitive contexts for morphological complexity: Dissociating inflection and derivation. *Language and Linguistics Compass, 4*(11), 1063-1073.
Brooks, T. L., & Cid de Garcia, D. (2015). Evidence for morphological composition in compound words using MEG. *Frontiers in human neuroscience, 9*, 215.
Butler, L. K., Pecukonis, M., Rogers, D. J., Boas, D. A., Tager-Flusberg, H., & Yücel, M. A. (2023). The role of the dorsolateral prefrontal cortex in the production and comprehension of phonologically and semantically related words. *Brain Sciences, 13*(7), 1113.
Cao, W. (2003). Investigating the differences between colloquial and literary words in modern Chinese [in Chinese]. *Yuyan jiaoxue yu yanjiu, 6*.
Carrasco-Ortiz, H., & Frenck-Mestre, C. (2014). Phonological and orthographic cues enhance the processing of inflectional morphology. ERP evidence from L1 and L2 French. *Front Psychol, 5*, 888. https://doi.org/10.3389/fpsyg.2014.00888
Cavalli, E., Colé, P., Badier, J.-M., Zielinski, C., Chanoine, V., & Ziegler, J. C. (2016). Spatiotemporal dynamics of morphological processing in visual word recognition. *Journal of Cognitive Neuroscience, 28*(8), 1228-1242.
Chung, K. K., Tong, X., Liu, P. D., McBride-Chang, C., & Meng, X. (2010). The processing of morphological structure information in Chinese coordinative compounds: an event-related potential study. *Brain Res, 1352*, 157-166. https://doi.org/10.1016/j.brainres.2010.06.069
Clahsen, H. (1999). Lexical entries and rules of language: A multidisciplinary study of German inflection. *Behavioral and brain sciences, 22*(6), 991-1013.
Collins, A. M., & Loftus, E. F. (1975). A spreading-activation theory of semantic processing. *Psychological review, 82*(6), 407.
Cui, L., Cong, F., Wang, J., Zhang, W., Zheng, Y., & Hyona, J. (2018). Effects of Grammatical Structure of Compound Words on Word Recognition in Chinese. *Front Psychol, 9*, 258. https://doi.org/10.3389/fpsyg.2018.00258
Dai, J. (1997). Syntactic, phonological, and morphological words in Chinese. *Trends in Linguistics Studies and Monographs, 105*, 103-134.
De Grauwe, S., Lemhöfer, K., Willems, R. M., & Schriefers, H. (2014). L2 speakers decompose morphologically complex verbs: fMRI evidence from priming of transparent derived verbs. *Frontiers in human neuroscience, 8*, 802.
Fiorentino, R., Naito-Billen, Y., Bost, J., & Fund-Reznicek, E. (2014). Electrophysiological evidence for the morpheme-based combinatoric processing of English compounds. *Cognitive neuropsychology, 31*(1-2), 123-146.
Frisson, S., Niswander-Klement, E., & Pollatsek, A. (2008). The role of semantic transparency in the processing of English compound words. *British Journal of Psychology, 99*(1), 87-107.
Frost, R., Kugler, T., Deutsch, A., & Forster, K. I. (2005). Orthographic structure versus morphological structure: Principles of lexical organization in a given language. *Journal of Experimental Psychology: Learning, Memory, and Cognition, 31*(6), 1293.
Grainger, J., & Beyersmann, E. (2020). Effects of lexicality and pseudo-morphological complexity on embedded word priming. *Journal of Experimental Psychology: Learning, Memory, and Cognition*.
Gu, J., Yu, L., & Ma, P. (2012). P2a reflects morphology structural processes in brain: Evidence from ERPs [in Chinese]. *Journal of Southwest University (Natural Science Edition), 34*(2), 137-144.
Hao, Y., Wu, C., & Duan, X. (2024). Processing mandarin Chinese compound words by native speakers and second language learners: word frequency, semantic transparency, and word structure. *Sage Open, 14*(2), 21582440241256249.
Hill, H., Strube, M., Roesch-Ely, D., & Weisbrod, M. (2002). Automatic vs. controlled processes in semantic priming—differentiation by event-related potentials. *International Journal of Psychophysiology, 44*(3), 197-218.
Hsu, C.-H., Pylkkänen, L., & Lee, C.-Y. (2019). Effects of morphological complexity in left temporal cortex: An MEG study of reading Chinese disyllabic words. *Journal of Neurolinguistics, 49*, 168-177. https://doi.org/10.1016/j.jneuroling.2018.06.004

Huang, B., & Liao, X. (2011). *Xiandai Hanyu* [in Chinese] (5 ed.). Higher Education Press.

Huang, C. Y., Lee, C. Y., Huang, H. W., & Chou, C. J. (2011a). Number of sense effects of Chinese disyllabic compounds in the two hemispheres. *Brain Lang, 119*(2), 99-109. https://doi.org/10.1016/j.bandl.2011.04.005

Huang, H. W., Lee, C. Y., Tsai, J. L., & Tzeng, O. J. (2011b). Sublexical ambiguity effect in reading Chinese disyllabic compounds. *Brain Lang, 117*(2), 77-87. https://doi.org/10.1016/j.bandl.2011.01.003

Ip, K. I., Hsu, L. S., Arredondo, M. M., Tardif, T., & Kovelman, I. (2017). Brain bases of morphological processing in Chinese-English bilingual children. *Dev Sci, 20*(5). https://doi.org/10.1111/desc.12449

Ip, K. I., Marks, R. A., Hsu, L. S., Desai, N., Kuan, J. L., Tardif, T., & Kovelman, L. (2019). Morphological processing in Chinese engages left temporal regions. *Brain Lang, 199*, 104696. https://doi.org/10.1016/j.bandl.2019.104696

Isel, F., Gunter, T. C., & Friederici, A. D. (2003). Prosody-assisted head-driven access to spoken German compounds. *Journal of Experimental Psychology: Learning, Memory, and Cognition, 29*(2), 277.

Jespersen, O. (1914). *A modern English grammar on historical principles* (Vol. 2). C. Winter.

Ji, H., & Gagné, C. L. (2007). Lexical and relational influences on the processing of Chinese modifier-noun compounds. *The Mental Lexicon, 2*(3), 387-417.

Jia, X., Wang, S., Zhang, B., & Zhang, J. X. (2013). Electrophysiological evidence for relation information activation in Chinese compound word comprehension. *Neuropsychologia, 51*(7), 1296-1301. https://doi.org/10.1016/j.neuropsychologia.2013.03.024

Jiang, N. (2018). *Second language processing: An introduction.* Routledge.

Laudanna, A., & Burani, C. (1995). Distributional properties of derivational affixes: Implications for processing. *Morphological aspects of language processing*, 345-364.

Lavric, A., Clapp, A., & Rastle, K. (2007). ERP evidence of morphological analysis from orthography: A masked priming study. *Journal of Cognitive Neuroscience, 19*(5), 866-877.

Lehtonen, M., Vorobyev, V. A., Hugdahl, K., Tuokkola, T., & Laine, M. (2006). Neural correlates of morphological decomposition in a morphologically rich language: An fMRI study. *Brain and Language, 98*(2), 182-193.

Leminen, A., Smolka, E., Dunabeitia, J. A., & Pliatsikas, C. (2019). Morphological processing in the brain: The good (inflection), the bad (derivation) and the ugly (compounding). *Cortex, 116*, 4-44. https://doi.org/10.1016/j.cortex.2018.08.016

Levelt, W. J. (1989). *Speaking: From intention to articulation.* MIT Press.

Libben, G. (2014). The nature of compounds: A psychocentric perspective. *Cognitive neuropsychology, 31*(1-2), 8-25.

Libben, G., Gagné, C. L., & Dressler, W. U. (2020). The representation and processing of compounds words. *Word Knowledge and Word Usage*, 336.

Libben, G., Gibson, M., Yoon, Y. B., & Sandra, D. (2003). Compound fracture: The role of semantic transparency and morphological headedness. *Brain and Language, 84*(1), 50-64.

Liu, D. (2016). The involvement of morphological information in the memorization of Chinese compound words: Evidence from memory errors. *Journal of psycholinguistic research, 45*(1), 157-176.

Liu, D. (2017). The influence of morphological structure information on the memorization of Chinese compound words. *Reading and Writing, 30*(8), 1813-1834.

Liu, D., & McBride-Chang, C. (2010a). Morphological processing of Chinese compounds from a grammatical view. *Applied Psycholinguistics, 31*(4), 605-617. https://doi.org/10.1017/s0142716410000159

Liu, D., & McBride-Chang, C. (2010b). What is morphological awareness? Tapping lexical compounding awareness in Chinese third graders. *Journal of Educational Psychology, 102*(1), 62-73. https://doi.org/10.1037/a0016933

Marelli, M., & Luzzatti, C. (2012). Frequency effects in the processing of Italian nominal compounds: Modulation of headedness and semantic transparency. *Journal of Memory and Language, 66*(4), 644-664.
Marslen-Wilson, W. D., & Tyler, L. K. (2007). Morphology, language and the brain: the decompositional substrate for language comprehension. *Philosophical Transactions of the Royal Society B: Biological Sciences, 362*(1481), 823-836.
Mitchell, R. L., Vidaki, K., & Lavidor, M. (2016). The role of left and right dorsolateral prefrontal cortex in semantic processing: A transcranial direct current stimulation study. *Neuropsychologia, 91*, 480-489.
Newman, A. J., Ullman, M. T., Pancheva, R., Waligura, D. L., & Neville, H. J. (2007). An ERP study of regular and irregular English past tense inflection. *Neuroimage, 34*(1), 435-445.
Olsen, S. (2012). Semantics of compounds. In C. Maienborn, K. von Heusinger, & P. Portner (Eds.), *Semantics: An international handbook of natural language meaning* (Vol. 3, pp. 2120–2150). Berlin: De Gruyter Mouton.
Pan, X., & Jared, D. (2020). Effects of Chinese word structure on object categorization in Chinese–English bilinguals. *Language and Cognition, 12*(3), 468-500. https://doi.org/10.1017/langcog.2020.8
Pion-Tonachini, L., Kreutz-Delgado, K., & Makeig, S. (2019). ICLabel: An automated electroencephalographic independent component classifier, dataset, and website. *Neuroimage, 198*, 181-197.
Schaeffer, J. D., Yennu, A. S., Gandy, K. C., Tian, F., Liu, H., & Park, H. (2014). An fNIRS investigation of associative recognition in the prefrontal cortex with a rapid event-related design. *J Neurosci Methods, 235*, 308-315. https://doi.org/10.1016/j.jneumeth.2014.07.011
Schiller, N. O. (2020). Neurolinguistic Approaches in Morphology. In *Oxford Research Encyclopedia of Linguistics*.
Silva, R., & Clahsen, H. (2008). Morphologically complex words in L1 and L2 processing: Evidence from masked priming experiments in English. *Bilingualism: Language and Cognition, 11*(2), 245-260.
Smolka, E., & Libben, G. (2017). 'Can you wash off the hogwash?'–semantic transparency of first and second constituents in the processing of German compounds. *Language, Cognition and Neuroscience, 32*(4), 514-531.
Su, B. (2016). The implicity, diversity and cognitive features of Chinese compounding [in Chinese]. *Xueshu Yanjiu* (1), 162-165.
Sun, J., Luo, X., & Pae, H. K. (2024). The Recognition of Chinese Compound Words by Native English-and Korean-speaking Learners of Chinese. *Reading in a Foreign Language, 36*(1), 1-22.
Sun, J., Pae, H. K., & Ai, H. (2021). The recognition of coordinative compound words by learners of Chinese as a foreign language: A mixed methods study. *Foreign Language Annals, 54*(4), 923-951.
Sun, J., Zhao, W., & Pae, H. K. (2020). Inter-character Orthographic Similarity Effects on the Recognition of Chinese Coordinative Compound Words. *J Psycholinguist Res, 49*(1), 125-145. https://doi.org/10.1007/s10936-019-09674-7
Taft, M. (1994). Interactive-activation as a framework for understanding morphological processing. *Language and Cognitive Processes, 9*(3), 271-294.
Treisman, A. M. (1960). Contextual cues in selective listening. *Quarterly Journal of Experimental Psychology, 12*(4), 242-248.
Tsang, Y.-K., Wong, A. W.-K., Huang, J., & Chen, H.-C. (2013). Morpho-orthographic and morpho-semantic processing in word recognition and production: evidence from ambiguous morphemes. *Language, Cognition and Neuroscience, 29*(5), 543-560. https://doi.org/10.1080/01690965.2013.790554
Tsang, Y. K., Huang, J., Lui, M., Xue, M., Chan, Y. F., Wang, S., & Chen, H. C. (2018). MELD-SCH: A megastudy of lexical decision in simplified Chinese. *Behav Res Methods, 50*(5), 1763-1777. https://doi.org/10.3758/s13428-017-0944-0

Wei, Y., Niu, Y., Taft, M., & Carreiras, M. (2023). Morphological decomposition in Chinese compound word recognition: Electrophysiological evidence. *Brain and Language, 241*, 105267.

Whiting, C. M., Marslen-Wilson, W. D., & Shtyrov, Y. (2013). Neural dynamics of inflectional and derivational processing in spoken word comprehension: laterality and automaticity. *Frontiers in human neuroscience, 7*, 759.

Wu, Y., Duan, R., Zhao, S., & Tsang, Y.-K. (2020). Processing Ambiguous Morphemes in Chinese Compound Word Recognition: Behavioral and ERP Evidence. *Neuroscience, 446*, 249-260.

Xu, Y., Graber, H. L., Schmitz, C. H., & Barbour, R. L. (2014). *nirsLAB: A Problem-Solving Environment for fNIRS Neuroimaging Data Analysis*.

Ye, J. C., Tak, S., Jang, K. E., Jung, J., & Jang, J. (2009). NIRS-SPM: statistical parametric mapping for near-infrared spectroscopy. *Neuroimage, 44*(2), 428-447.

Zhang, B., & Peng, D. (1992). Decomposed storage in the Chinese lexicon. In *Advances in Psychology* (Vol. 90, pp. 131-149). Elsevier.

Zhang, P., Zhang, Q., Peng, G., Song, X., & Bai, X. (2016). The Representation of Chinese Two-character Words in Mental Lexicon: Evidence from an fNIRS Study [in Chinese]. *Psychological Science, 39*(4), 849-855.

Zou, L., Packard, J. L., Xia, Z., Liu, Y., & Shu, H. (2015). Neural Correlates of Morphological Processing: Evidence from Chinese. *Front Hum Neurosci, 9*, 714. https://doi.org/10.3389/fnhum.2015.00714

Zou, L., Packard, J. L., Xia, Z., Liu, Y., & Shu, H. (2019). Morphological and Whole-Word Semantic Processing Are Distinct: Event Related Potentials Evidence From Spoken Word Recognition in Chinese. *Front Hum Neurosci, 13*, 133. https://doi.org/10.3389/fnhum.2019.00133

Chapter 8
Neural Correlates of Morphological Processing in the Chinese–English Bilingual Brain

8.1 Introduction

Building on the findings from Chaps. 6 and 7, which examined the cognitive and neural bases of morphological processing in Chinese native speakers, this chapter extends the investigation to the bilingual context. While previous chapters demonstrated that Chinese readers recruit primarily the left frontal brain regions to encode morphological constraints and structural differences, a key question remains as to whether such mechanisms are language-specific or shared across languages. In bilingual contexts, especially for learners whose L1 and L2 differ typologically, morphological processing involves not only the representation of morphemes within each language but also the potential transfer and interaction between the two systems. In particular, the extent to which Chinese and English share cognitive and neural resources for morphology, and whether experience in one language facilitates or constrains processing in the other, is central to understanding cross-linguistic transfer and bilingual learning. This chapter therefore investigates the neural correlates of morphological processing in the Chinese–English bilingual brain, with particular attention to the extent of overlap and divergence in morphological representation across the two languages.

8.2 Theoretical Frameworks and Behavioral Evidence of Cross-Linguistic Transfer

In the context of globalization, bilingualism and multilingualism have become increasingly prevalent, making the question of how linguistic knowledge transfers across languages both theoretically and practically important. For L2 learners, an especially critical issue concerns whether morphological knowledge acquired in the

F. Gao, *The When, Where, and How of Morphological Processing in the Human Brain*, https://doi.org/10.1007/978-981-95-5981-7_8

L1 can be transferred to facilitate L2 learning. As discussed earlier in Sect. 3.5.3, morphological knowledge can be broadly divided into explicit morphological awareness and implicit morphological processing, both of which have demonstrated potential for transfer. However, the extent and nature of such transfer are often shaped by multiple factors, including the typological distance between L1 and L2 as well as individual learner variables.

In the domain of morphological awareness, the transfer facilitation model (Koda 2008) provides a framework for explaining cross-linguistic sharing in bilingual contexts. Recent meta-analyses have shed systematic light on this issue. Ke et al. (2021), synthesizing 34 studies, reported small but significant correlations between L1 and L2 morphological awareness ($r = 0.30$), and between L1 morphological awareness and L2 decoding ($r = 0.35$) as well as L2 reading comprehension ($r = 0.39$). Interestingly, these effects were not significantly moderated by age or script differences between L1 and L2, suggesting that a portion of morphological awareness may indeed be a transferable cognitive resource across languages. In contrast, Cheng et al. (2025) conducted a much larger meta-analysis of 109 studies and revealed a more nuanced picture. Among monolinguals, the correlation between morphological awareness and vocabulary knowledge was moderate ($r = 0.43$), whereas among bilinguals, the correlation between L1 and L2 morphological awareness was $r = 0.33$, and between L1 morphological awareness and L2 vocabulary was $r = 0.26$. These findings indicate that cross-linguistic transfer of morphological awareness is real but of modest strength. Importantly, age and language distance were found to moderate the effects. As learners grew older, the correlation between L1 and L2 morphological awareness tended to weaken. Moreover, bilinguals with typologically closer language pairs (e.g., Spanish–English) showed stronger transfer effects than those with more distant pairs (e.g., Chinese–English) (X. Sun et al. 2022). Additionally, task format and assessment type also played a role, with production-based morphological awareness tasks generally yielding stronger correlations, and derivational and compounding awareness showing stronger links to vocabulary growth than inflectional awareness.

On the other hand, research in morphological processing has been more supportive of transfer effects, especially with respect to the role of language distance. For example, J. Sun et al. (2024) compared English- and Korean-speaking learners of L2 Chinese and found that both groups showed a similar hierarchy of difficulty in processing Chinese compounds: subject–predicate structures were the hardest to recognize, followed by verb–complement structures. However, the limited number of items per structure type reduced the explanatory power of this study. Additionally, although the authors suggested that subject–predicate compounds are a feature unique to Chinese, English also contains comparable formations (e.g., *sunrise* and *sunset*) that parallel subject–predicate structures. Similarly, Hao et al. (2024) examined Chinese learners' processing of compound structures but did not strictly control for L1–L2 structural similarity, which restricted its contribution to clarifying transfer mechanisms. Taken together, these behavioral studies suggest the plausibility of cross-linguistic transfer in morphological processing, yet the evidence remains scattered and constrained by methodological limitations.

In the broader field of bilingual processing, two influential theoretical frameworks have been advanced to explain transfer of processing strategies. The Unified Competition Model (MacWhinney and Kroll 2005) posits that when L1 and L2 share overlapping structures, learners are likely to draw on common cognitive resources and strategies, facilitating transfer. By contrast, the Shallow Structure Hypothesis (Clahsen and Felser 2006, 2018) argues that L2 processing relies more on surface-level lexical cues than on deep structural parsing, suggesting more limited transfer. Although these models were primarily developed for syntactic processing, findings from earlier chapters of this book highlight the relative independence of morphology from syntax and semantics. This raises a critical question: Can morphological processing itself, as a cognitive strategy and resource, be transferred from L1 to L2, and how is such transfer moderated by language structures? Addressing this question requires going beyond behavioral paradigms and turning to neuroscientific evidence, which provides a finer-grained perspective on the cognitive and neural correlates of transfer effect.

8.3 Neuroimaging Evidence of Morphological Processing in Bilinguals

Drawing on electrophysiological and neuroimaging evidence, existing research has established that morphological processing constitutes a dissociable component of the human mental lexicon, distinct from semantic processing (e.g., Bölte et al. 2009; Zou et al. 2019). ERP studies on both alphabetic and logographic languages suggested that morphological parsing engages multiple temporal stages. For example, an LAN effect, often associated with morpho-syntactic violation detection (Morris and Holcomb 2005; Rossi et al. 2006), was observed when participants processed morphologically manipulated German adjectives (Bölte et al. 2009). Similarly, in Chinese, Chap. 6 found that early and late N400 components have been linked to semantic and morphological constraints, respectively, when participants processed real compounds versus illegal nonwords. These findings suggest that morphological processing may recruit a relatively late, top-down mechanism. In contrast, earlier components such as the P/N250 have been linked to rapid, form-based morphological decomposition in masked priming paradigms with very short SOAs (K. K. Chung et al. 2010; Gu et al. 2012; Morris and Stockall 2012).

Converging results from MEG and fMRI studies further highlight the critical involvement of the left frontal and temporal cortices in morphological processing across languages (Bakker et al. 2013; Bick et al. 2008; Forgács et al. 2012; Fruchter et al. 2013; Lehtonen et al. 2006; Leminen et al. 2019; Liu et al. 2013; Whiting et al. 2013; Zhao et al. 2021). Within this network, the left IFG appears to play a language-general role in morphological analysis, sometimes in cooperation with subcortical regions such as the basal ganglia and cerebellum (Leminen et al. 2019; Zou et al. 2015). Chinese studies add an important nuance, showing that the superior and middle temporal gyri (STG, MTG) are particularly engaged during morphological as opposed to phonological tasks (Ip et al. 2019), likely reflecting the morpho-syllabic salience of Chinese. MEG evidence further indicates that the left posterior temporal cortex is modulated by morphological complexity during Chinese compound word reading (Hsu et al. 2019). Collectively, these findings suggest both universal and language-specific neural mechanisms underlying morphological processing, providing a foundation for examining how such processes operate in bilingual contexts.

So far, however, relatively little is known about the neural basis of morphological processing in bilingual brains. With bilingualism becoming increasingly prevalent in a globalized world, probing how morphology is processed and learned across two languages offers valuable implications for biliteracy development and pedagogical practice. A key question is whether bilingual individuals rely on similar morphological strategies when processing their L1 and L2. Existing studies have primarily focused on comparing bilinguals' L2 performance with that of native speakers of the same language (e.g., W. Chung et al. 2019b; Lehtonen et al. 2009; Marks et al. 2021; Prehn et al. 2018), while the direct interplay between L1 and L2 systems within the same bilingual brain remains underexplored. One major reason might be the confounding effect of typological distance between the two languages. To address this, some studies have examined highly proficient bilinguals or targeted language structures shared across L1 and L2. For example, Lehtonen et al. (2009) reported distinct neural signatures when Finnish-Swedish bilinguals engaged in morphological processing during fMRI scans. L1 Finnish triggered robust activations in the left IFG and posterior temporal areas, whereas these regions were not involved in L2 Swedish, pointing to divergent processing strategies.

At the same time, accumulating evidence also supports the idea that L2 morpho-syntactic processing may draw upon neural resources already established for L1, particularly when the two languages share structural similarities. This view resonates with the Unified Competition Model (MacWhinney and Kroll 2005), which posits that overlapping linguistic structures in L1 and L2 recruit common cognitive mechanisms. For instance, similar fronto-temporal networks have been observed for morpho-syntactic processing in both L1 and L2 (e.g., Vingerhoets et al. 2003; Wattendorf et al. 2001). Nonetheless, bilingual morphological processing is not entirely uniform: proficiency, AoA, and structural divergence modulate the extent to which L2 converges with or diverges from L1 neural patterns. Taken together, the extant evidence tends to indicate that bilingual morphological processing is shaped by both shared and language-specific mechanisms.

Given these nuanced findings, recent perspectives underscore that cross-language transfer of literacy skills, including morphological knowledge, should be conceived as a dynamic and interactive process. S. C. Chung et al. (2019a) proposed an interactive framework to capture the complexity of bilingual development, arguing that transfer is shaped by the interplay of multiple factors, such as structural similarities between the two languages, the specific literacy sub-skill, learner proficiency, and contextual influences. With respect to morphological processing, the degree to which bilinguals' L2 neural patterns converge with or diverge from those of their L1 remains inconclusive, especially for bilinguals whose two languages are typologically distant. As reviewed in Chap. 3, Chinese relies primarily on compounding as its dominant morphological strategy, whereas English is characterized by extensive use of inflection and derivation. Despite the importance of this contrast, neuroimaging studies focusing on Chinese–English bilinguals are still relatively scarce, with only a few recent investigations targeting bilingual children.

One such study by Ip et al. (2017) examined the neural correlates of morphological processing in Chinese–English bilingual children (aged 6–13) compared to English monolinguals using an auditory morphological structure judgment task. Results revealed that, relative to monolingual peers, bilingual children exhibited heightened activation in the left MTG during English processing, a region associated with lexico-semantic integration and L2 literacy development. Within the bilingual group, L2 English morphology also elicited stronger engagement of the left IFG compared to L1 Chinese, suggesting the recruitment of both language-general and language-specific neural resources. The authors interpreted the IFG modulation as an index of bilinguals' sensitivity to the distinct morphological constraints of Chinese and English. Nevertheless, the use of different morphological structures across the two languages (English derivation vs. Chinese compound) may have confounded the interpretation of transfer effects, which merits further verification.

Complementary evidence from X. Sun et al. (2023) compared how bilingual children from typologically distant (Chinese–English) and closer (Spanish–English) language backgrounds process English morphology relative to monolingual peers. Both bilingual groups showed a similar effect. Higher proficiency in the home language was linked to stronger activation in the left STG when they processed English word structures that differed most from their home language. Compared with Chinese–English bilinguals, who showed stronger frontal activation for English derivations but no enhancement for compounding due to cross-language differences, Spanish–English bilinguals exhibited reduced frontal activation for derivation and heightened temporal activation for compounding, reflecting their greater lexical overlap with English. Together, these studies highlight that bilingual morphological processing is jointly shaped by universal brain networks and language-specific adaptations, with cross-language transfer constrained by both structural distance and the literacy domain under investigation.

Previous neuroimaging studies on bilingual morphological processing have predominantly focused on children and have employed auditory tasks (Ip et al. 2017; X. Sun et al. 2023). However, the auditory modality may not fully capture the salience of morphology, which in many cases relies on print-to-meaning mapping. Moreover, developmental studies cannot directly address how bilingual morphological processing unfolds beyond the critical period, nor how multiple interacting factors shape bilingual language transfer in adulthood. To extend this line of inquiry, the present study investigates adult Chinese–English bilinguals, thereby providing insights into morphological processing after the critical period and offering a richer understanding of the dynamic interplay between L1 and L2 in bilingual brains.

Specifically, the objective of this study is to examine the shared and distinct neural correlates of L1 and L2 morphological processing by employing a visual morphological priming paradigm with derivational morphology, a structure common to both Chinese and English. To capture the complementary temporal and spatial dimensions of neural activity, both EEG and fNIRS will be recorded simultaneously while participants perform morphological tasks. By focusing on a shared morphological structure, the design minimizes confounds arising from structural differences between the two languages. We hypothesize that the observed spatio-temporal activation patterns will primarily reflect the processing strategies employed for L1 and L2 morphology, rather than mere structural divergence, thereby contributing to a more precise characterization of cross-language transfer in adult bilinguals.

8.4 Methods

8.4.1 Participants

Thirty native speakers of Mandarin Chinese (15 females; $M = 22.2$ years, $SD = 3.2$,) were recruited from the University of Macau and received monetary compensation for participation. All were right-handed, had normal or corrected-to-normal vision, and reported no history of neurological or psychiatric disorders. On average, they began learning English as a foreign language at 7.4 years of age ($SD = 2.2$) and none had resided in an English-speaking country for longer than six months. Before taking part in the main experiment, participants completed the online LexTALE test (Lemhöfer and Broersma 2012) to assess their English vocabulary knowledge. Their mean score was 57.21 out of 100 ($SD = 6.74$), which corresponds to an intermediate level (with 60% commonly used as the threshold for medium proficiency). The study protocol was reviewed and approved by the Institutional Review Board of the University of Macau, and informed consent was obtained from all participants prior to testing.

8.4.2 Materials

In the Chinese condition, 60 high-frequency derivational words (mean log frequency = 1.65), selected from the SUBTLEX-CH corpus (Cai and Brysbaert 2010), were used as targets. These derivational forms involved six (semi-)prefixes (e.g., 非- /fei1/, "non-") and forty (semi-)suffixes (e.g., -师/shi1/, "-professional"), identified on the basis of a comprehensive summary of Chinese affixes and semi-affixes (Yang 2020). Within the morphological priming paradigm, each target was preceded by its root morpheme. For example, the target 书法家/shu1 fa3 jia1/ ("calligrapher") was primed by 书法/shu1 fa3/ ("calligraphy"). To provide a semantic control condition, 60 Chinese compound words were paired with semantically related primes, which were retrieved from the Small World of Words Project (Li et al. 2024). These prime–target pairs were matched in meaning but shared no morphological, orthographic, or phonological overlap.

For the English condition, 60 suffixed words (e.g., containing -er, -or, -ness, -ment) were selected from the SUBTLEX-US corpus (Brysbaert and New 2009). In parallel, 60 semantically related prime–target pairs were drawn from the University of South Florida Free Association Norms dataset (Nelson et al. 2004) to serve as the semantic control set.

A semantic judgment task was designed that included both Chinese and English stimuli. Participants were instructed to decide whether the meaning of each target word referred to a person or not. In each of the four conditions (CHMP-Chinese morphological priming, CHSP-Chinese semantic priming, ENMP-English morphological priming, and ENSP-English semantic priming), half of the target words denoted people (e.g., 笑星/xiao4 xing1/, "comedian"; farmer), and the remaining half did not (e.g., 比赛/bi3 sai4/, "competition"; piano). To ensure validity, seven university students with different levels of English proficiency evaluated the people-relatedness of all target words, and their offline judgments showed complete consistency. In addition, eight campus volunteers rated the familiarity of the 120 target words on a 1–9 scale, yielding high familiarity scores (all above 8.2) for both Chinese and English items. Another nine raters assessed the semantic association between primes and targets across conditions, confirming strong relatedness. Key lexical variables, including log frequency, word length, stroke number, familiarity, and semantic association, were carefully controlled and matched across conditions (see Table 8.1). Chinese and English stimuli were provided in Appendices E and F, respectively.

8.4.3 Procedure

Participants completed a semantic judgment task implemented in a block design, which is particularly advantageous for fNIRS because it provides a more stable and robust hemodynamic response compared to event-related paradigms (e.g., a slow event-related design in Chap. 6 and a rapid event-related design in Chap. 7), thereby improving the signal-to-noise ratio and reliability of activation patterns. The experiment consisted of three runs, each containing eight blocks (two blocks per condition per run). Each block included ten trials, with five targets denoting people and five targets not related to people, which were presented in a random order. Importantly, the "people-related or not" decision was to minimize participants' potential response strategies, as the task required them to process the meaning of the target rather than relying on superficial features.

Each block was separated by a central fixation cross displayed for four seconds. A trial began with a prime shown for 150 ms, followed by a 50-ms mask of two red asterisks. The target word then appeared and remained visible until the participant responded or until 3 seconds had elapsed, after which a blank screen was presented

Table 8.1 Lexical characteristics and examples of primes and targets across conditions

Language	Priming Type	Semantic Judgment	Prime				Target					Semantic Relatedness
			Example	Frequency	Word Length	Stroke Number	Example	Frequency	Word Length	Stroke Number	Familiarity	
Chinese (L1)	Morphological priming	People-related	笑/ xiao4/, laugh	2.9 (1.1)	1.7 (0.5)	13.9 (6.8)	笑星/xiao4 xing1/, comedian	1.9 (0.8)	2.7 (0.5)	20.2 (7)	8.3 (0.5)	7.0 (0.6)
		Non people-related	现代/ xian4 dai4/, modern	2.9 (0.8)	1.7 (0.5)	13 (5.3)	现代化/xian4 dai4 hua4/, modernization	1.4 (0.9)	2.7 (0.5)	20.7 (6.4)	8.2 (0.6)	6.8 (0.9)
	Semantic priming	People-related	监狱/ jian1 yu4/, prison	2.9 (0.9)	1.8 (0.4)	13 (3.7)	罪犯/zui4 fan4/, criminal	2.6 (1.2)	2.2 (0.4)	17.6 (6.3)	8.3 (0.5)	6.8 (0.7)
		Non people-related	游戏/ you2 xi4/, game	3 (1.1)	1.7 (0.4)	14.9 (4.5)	比赛/bi3 sai4/, competition	2.9 (1.2)	2.2 (0.4)	16.6 (4.7)	8.4 (0.5)	6.7 (0.7)
English (L2)	Morphological priming	People-related	farm	3.3 (0.6)	5.8 (1.8)	/	farmer	2.5 (0.7)	8.1 (1.6)	/	8.6 (0.5)	7.1 (0.5)
		Non people-related	slow	3.5 (0.6)	5.4 (1.6)	/	slowness	2.7 (0.7)	8.3 (1.9)	/	8.6 (0.7)	7.2 (0.6)
	Semantic priming	People-related	hospital	3.6 (0.5)	5 (1.5)	/	nurse	3.4 (0.5)	5.7 (2)	/	8.6 (0.6)	6.5 (0.7)
		Non people-related	music	3.9 (0.6)	5.3 (1.5)	/	piano	3.6 (0.6)	6.1 (1.7)	/	8.7 (0.5)	6.5 (0.7)

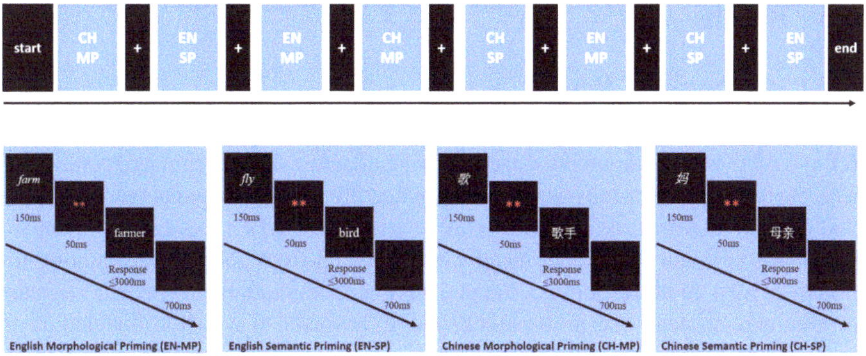

Fig. 8.1 Schematic illustration of the experimental procedure

for 700 ms. Responses were made via button press on a keyboard. The overall experimental sequence is illustrated in Fig. 8.1.

8.4.4 EEG-fNIRS Data Acquisition and Analysis

During the lexical decision task, EEG and fNIRS signals were collected simultaneously. EEG signals were recorded using the BrainVision actiChamp system with 32 active electrodes, sampled at 500 Hz and referenced online to the left mastoid. Electrode impedances were kept below 25 kΩ throughout acquisition. Data preprocessing was carried out in EEGLAB. Continuous data were re-referenced to the common average and filtered with a 0.01–30 Hz bandpass. Epochs were segmented from −200 ms pre-target onset to 800 ms post target. ICA was applied to remove ocular and line noise artifacts. Trials with amplitudes exceeding ±100 μV were excluded, and electrodes showing persistent artifacts after correction were interpolated using spherical spline averaging.

The fNIRS data were obtained using the NIRScout system, which employs dual-wavelength light (760 nm and 850 nm) at a sampling frequency of 7.81 Hz. A montage of eight sources and eight detectors formed 22 measurement channels, covering the frontal and temporal regions of the left hemisphere, with source–detector distances fixed at 3 cm. The stereotactic positions of these channels were mapped to MNI coordinates following the international 10/20 system, and anatomical information (Brodmann areas, cortical labels, overlap percentages) was derived using the NIRS_SPM toolbox (Ye et al. 2009). Offline preprocessing was conducted with nirsLAB (Xu et al. 2014): signals were band-pass filtered between 0.01–0.2 Hz, discontinuities and spike artifacts were removed using built-in correction algorithms, and hemodynamic responses were calculated according to the Beer–Lambert law. HbO changes were then submitted to a GLM analysis with a canonical HRF, yielding beta estimates for each of the four experimental conditions.

8.5 Results

8.5.1 Behavioral Results

RT and ACC were computed for each participant across all experimental conditions. The overall mean accuracy was 94.9%, indicating that participants were attentive and compliant with task demands. Outlier trials with RTs exceeding ±3 standard deviations from the individual mean were excluded from analysis, accounting for less than 1.9% of all responses. The people relatedness judgment was incorporated to minimize strategy use in the block design; however, it was not modeled as an independent factor. Cleaned datasets were analyzed with linear mixed-effects models using the lme4 package and ANOVA function in R (Bates et al. 2014; Kuznetsova et al. 2015), with language and priming type as fixed effects, and people-relatedness and participant as random factors.

RT analyses revealed a significant main effect of language (estimate = 130.99, SE = 7.16, t = 18.30, $p < 0.001$): Chinese items (659 ± 114 ms) were recognized more quickly than English items (776 ± 137 ms). Priming type also yielded a significant effect (estimate = 27.17, SE = 7.09, t = 3.83, $p < 0.001$), with morphological priming (710 ± 131 ms) producing shorter RTs than semantic controls (724 ± 119 ms). Furthermore, a language × priming type interaction was observed (estimate = −28.96, SE = 10.09, t = −2.87, $p < 0.01$). Post hoc comparisons showed that morphological priming facilitated faster responses compared to semantic priming in L1 Chinese ($p = 0.0001$), but no such effect was found in L2 English ($p = 0.803$).

Accuracy analyses showed that Chinese targets (96.73% ± 0.30%) were recognized more accurately than English targets (93.67% ± 0.41%), estimate = 0.705, SE = 0.117, z = 6.03, $p < 0.0001$. No significant priming effect ($p = 0.580$) or interaction effect ($p = 0.814$) was detected for accuracy. Behavioral results were visualized in Fig. 8.2.

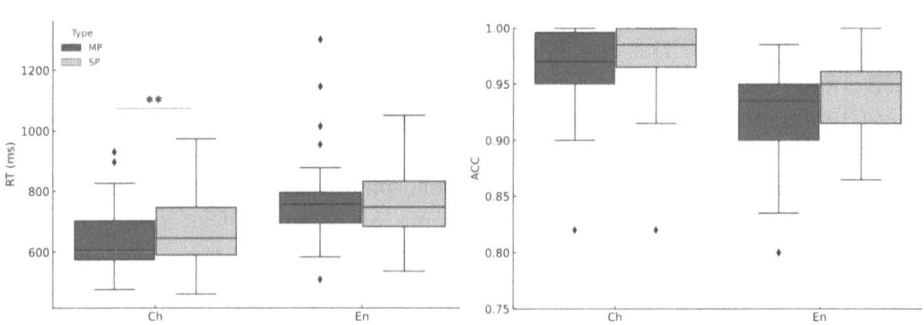

Fig. 8.2 RT and ACC results across the four conditions

8.5.2 ERP Results

Based on the grand-average waveforms (Fig. 8.3), the ERP data revealed an early left-anterior negativity (ELAN) shortly after target onset, followed by a positivity at 200–300 ms (P250), and a later left-lateralized negativity between 300–500 ms. To assess these effects, a repeated-measures ANOVA was conducted with four within-subject factors: language (L1 Chinese vs. L2 English), priming type (morphological vs. semantic), laterality (left: AFF5h, FC1, CPP5h, CP1, P3; midline: FCz, Pz; right: FC2, FC6, P4, CP2, CP6), and region (anterior: AFF5h, FC1, FCz, FC2, FC6; posterior: CPP5h, CP1, P3, P4, CP2, CP6).

First, the scalp distribution from 80 to 150 ms (see Fig. 8.4) showed a clear early negativity, consistent with an ELAN effect. A significant main effect of language was observed, $F\ (1{,}29) = 41.362$, $p < 0.001$, $\eta^2_p = .588$, with L1 Chinese ($-1.20 \pm 0.23\ \mu V$) eliciting stronger negativity than L2 English ($-0.52 \pm 0.21\ \mu V$). In addition, anterior electrodes produced more pronounced negativities ($-3.28 \pm 0.42\ \mu V$) than posterior sites ($1.57 \pm 0.25\ \mu V$; $p < 0.001$), and midline electrodes were more negative than lateral ones (*ps* < 0.001). A hemispheric trend was also noted, with stronger negativity in the left ($-0.53 \pm 0.26\ \mu V$) than in the right hemisphere ($-0.44 \pm 0.20\ \mu V$), although this difference did not reach significance. The three-way interaction of region × language × priming type reached marginal significance, $F\ (1{,}29) = 3.849$, $p = 0.059$, $\eta^2_p = .117$. Follow-up tests showed that in Chinese, morphological priming elicited significantly greater negativity in anterior regions compared to semantic priming ($-3.80 \pm 0.51\ \mu V$ vs. $-3.14 \pm 0.44\ \mu V$; $p = 0.03$). By contrast, no morphological priming effect was found in English ($p = 0.61$). These findings indicate that in the ELAN time window, early

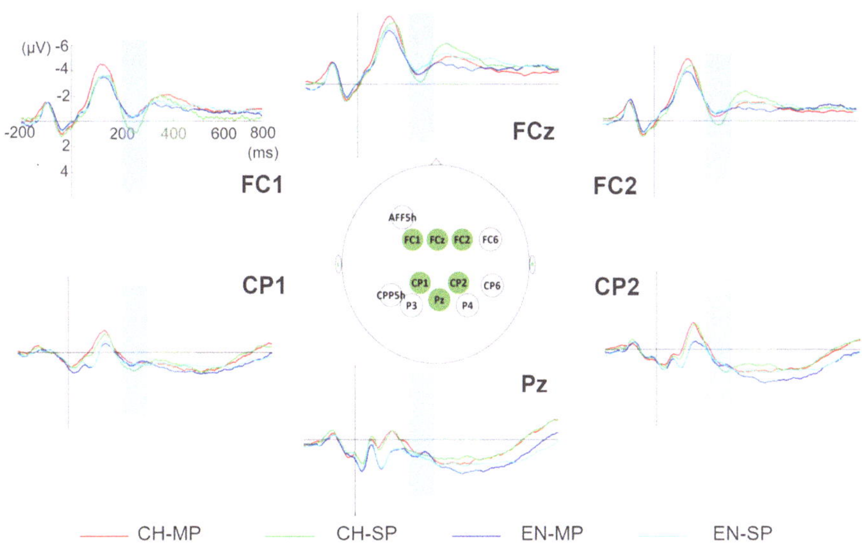

Fig. 8.3 Grand-average ERP brainwaves from representative electrodes

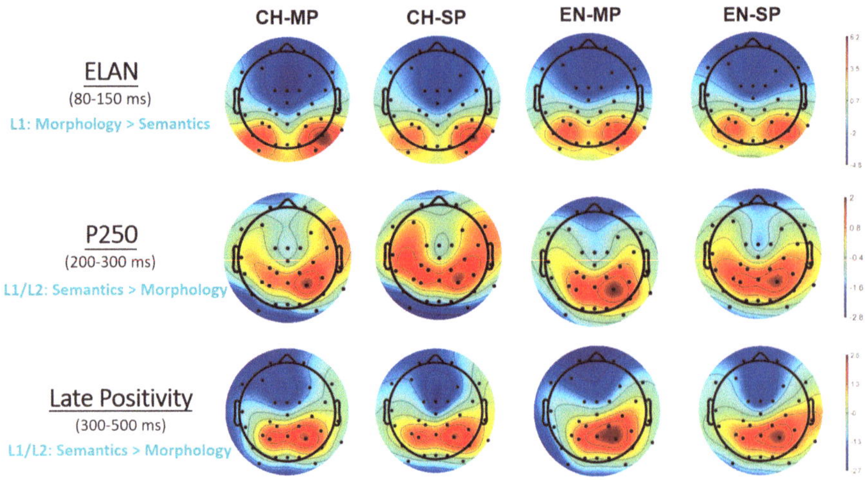

Fig. 8.4 Brain topographies of the three selected ERP components and main findings

morphological parsing is robust for the L1 but attenuated for the L2, with effects most pronounced at anterior and midline sites.

Then, P250 analyses showed a robust main effect of region, F (1,29) = 10.631, p = 0.003, η^2_p = 0.263, with posterior electrodes exhibiting stronger positivity (1.14 ± 0.25 μV) than anterior electrodes (−0.53 ± 0.40 μV). Moreover, the region × language interaction reached significance, F (1,29) = 4.997, p = 0.033, η^2_p = .147. Chinese tasks were associated with enhanced anterior negativity compared to English tasks (p = 0.048). A significant laterality × priming type interaction also emerged, F (2,58) = 5.597, p = 0.011, η^2_p = 0.162, such that semantic priming produced greater positivity than morphological priming in the left hemisphere (p = 0.005). No additional main effects or interactions involving language or priming type were observed.

Within the late time window of 300-500 ms, negativity was significantly stronger over the frontal sites (−1.37 ± 0.38 μV) compared to posterior electrodes (1.70 ± 0.24 μV), and was more pronounced in the left hemisphere (−0.08 ± 0.25 μV) relative to the right (0.63 ± 0.21 μV; ps < 0.05). A significant three-way interaction among region, laterality, and priming type was also observed, F (2,58) = 9.99, p = 0.002, η^2_p = .256. Follow-up analyses revealed that in the midline fronto-central cortex, semantic priming (−2.33 ± 0.44 μV) elicited stronger negativity than morphological priming (−1.61 ± 0.53 μV; p = 0.013).

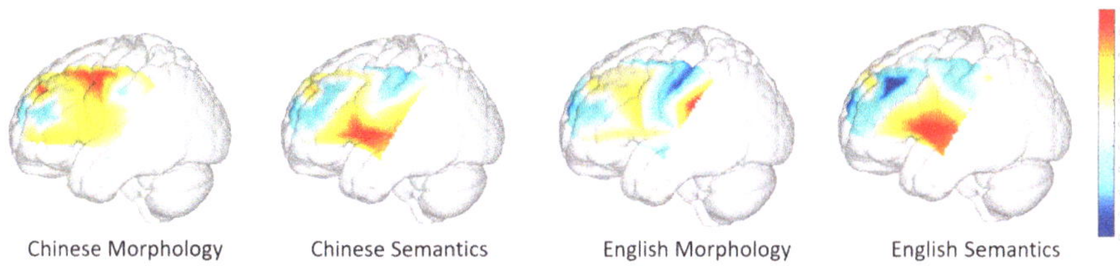

Fig. 8.5 Brain activation patterns associated with the four conditions

8.5.3 *fNIRS Results*

A series of two-way repeated-measures ANOVAs with factors language and priming type were carried out on beta values for each channel. Significant language main effects were detected at channels 1, 3, and 19 (the left frontopolar regions), where Chinese tasks elicited stronger hemodynamic responses than English tasks ($ps < 0.05$). In addition, three channels showed reliable priming type effects with divergent patterns. Specifically, morphological priming induced greater activation than semantic priming in channel 9 (DLPFC, $p < 0.05$), whereas semantic priming produced stronger responses in channels 15 and 16 (STG and MTG, $ps < 0.01$). Brain activation patterns across the four conditions are visualized in Fig. 8.5.

To further examine priming-related and cross-linguistic effects, four sets of pairwise comparisons were conducted. Within Chinese tasks, semantic priming yielded greater activation than morphological priming in the STG and MTG (channels 15 and 16). For English, morphological priming led to enhanced activation in the DLPFC (channel 9), while semantic priming recruited stronger responses in the STG and MTG (channels 15 and 16). Regarding cross-language contrasts, Chinese morphological priming was associated with more pronounced activity in the frontopolar cortex (channels 1, 3, 4) relative to English. Similarly, Chinese semantic priming evoked stronger activation in the frontopolar area (channel 3) compared to its English counterpart. Comparison results were demonstrated in Fig. 8.6.

8.6 Discussion

Using derivational morphology as a shared testing ground, this study investigated how cross-language similarity shapes the temporal and spatial neural mechanisms underlying L1 and L2 morphological processing in Chinese–English adult bilinguals. Behaviorally, participants demonstrated reliable language effects and morphological priming facilitation, with faster responses and higher accuracy in Chinese than in English. Importantly, morphological facilitation was largely driven by L1 Chinese, since no priming advantage was found in English. This asymmetry suggests that bilinguals may employ sub-lexical parsing strategies in their native language, whereas L2 morphological processing remains less sensitive and automatic

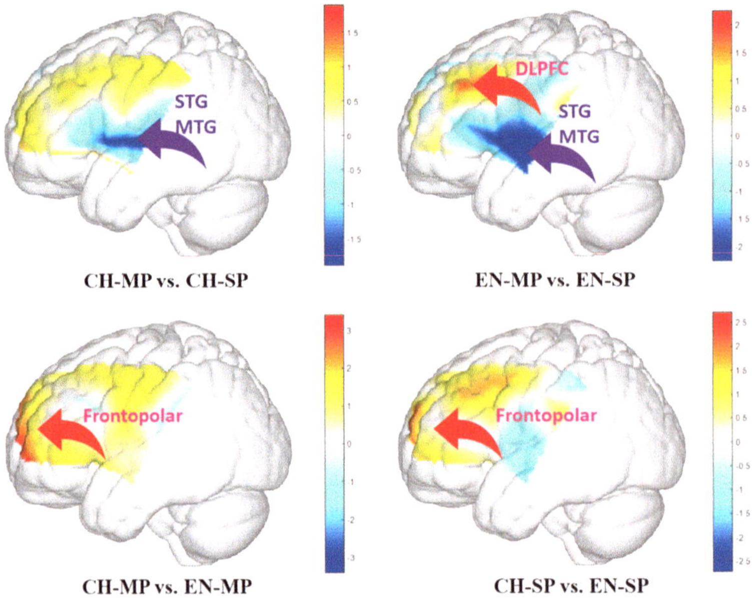

Fig. 8.6 Brain activation difference among the four pairwise comparisons

at this grain size. Chapter 5 demonstrated a preference for whole-word representation in compounds, as indicated by stronger repetition priming at the whole-word level compared to the morpheme level. Nevertheless, both whole-word and morphemic units are integral to the organization of the mental lexicon. It is also plausible that L1 Chinese relies more heavily on a decomposition strategy for derivational forms, given that the psycholinguistic mechanisms underlying Chinese derivation have received limited attention to date. The subsequent ERP and fNIRS analyses were therefore designed to determine whether L2 morphological processing converges with or diverges from L1 patterns at the neural level.

Our fNIRS analyses highlighted the left prefrontal cortex as a central hub for bilingual morphological processing. ANOVA results showed that morphological priming elicited stronger hemodynamic responses in the DLPFC compared to semantic priming. Pairwise tests further indicated that this effect was mainly driven by L2 English, while no reliable difference was observed between Chinese morphological and semantic priming. By contrast, purely semantic associations between primes and targets produced stronger activation in the left temporal cortex, including the STG and MTG, than morphological relations. This dissociation suggests that bilinguals engage distinct neural mechanisms for encoding morphology versus semantics. Specifically, the left frontal cortex appears to serve as a language-general region for morphological processing, a finding in line with our previous chapters and earlier work using varied tasks and modalities (Leminen et al. 2019; Schiller 2020; Zou et al. 2015). At the same time, the left temporal cortex remains central to differentiating semantic from morphological relationships, with semantic processing often outweighing morphological parsing. Supporting this view, Hsu et al. (2019) reported stronger temporal activation when reading morphologically complex Chinese compounds compared to monomorphemic words, and Ip et al. (2019) observed significant STG and MTG activation during a Chinese morphological construction task, linked to phonological and semantic analyses. Since both studies did not fully exclude semantic influence, it is plausible that temporal activations reflect semantic processing rather than purely morphological parsing. Consequently, activity in the left prefrontal cortex may provide the most robust marker of shared morphological processing in Chinese–English bilinguals.

A key contribution of the present study lies in documenting cross-linguistic contrasts within the frontal cortex. L1 Chinese consistently elicited stronger frontopolar activation than L2 English, regardless of priming type. This pattern is broadly consistent with monolingual neuroimaging research showing heightened frontal engagement for Chinese reading (Tan et al. 2005; Tan et al. 2003), but differs from findings in bilingual children (Ip et al. 2017), who exhibited stronger L2 frontal activations. We interpret this divergence in light of age of acquisition and bilingual profile: unlike early bilingual children immersed in English, the present adult participants were late bilinguals with stronger L1 dominance. By capitalizing on derivational morphology present in both languages, our study further showed that bilingual adults do not recruit distinct neural substrates for L1 versus L2 morphology, but instead vary in the degree of engagement of shared frontal regions, reflecting native proficiency and morphological competence.

The present findings also provide a useful contrast to X. Sun et al. (2023), who examined English morphological processing in Chinese–English and Spanish–English bilingual children using fNIRS. Sun and colleagues reported both common bilingual effects and language-specific differences. Across groups, higher home-language proficiency predicted greater left STG activation when processing English structures most dissimilar from the home language. At the same time, Chinese bilinguals showed stronger left frontal activation during English derivational tasks, reflecting the relatively lower frequency of such structures in Chinese, whereas Spanish bilinguals displayed relatively greater temporal engagement for compounding, consistent with Spanish–English structural mismatches. Our study differs in two key respects. First, we targeted adult late bilinguals rather than children in a critical developmental window, thereby reducing the confounding effect of neural plasticity associated with early dual-language exposure. Second, by employing a visual priming paradigm with shared derivational morphology across Chinese and English, we minimized structural mismatches and semantic confounds. Under these conditions, we observed largely overlapping networks for L1 and L2, but with quantitative rather than qualitative differences.

Interestingly, the ERP results also revealed both shared and distinct electrophysiological signatures of bilingual morphological processing. At the earliest stage, both L1 and L2 waveforms exhibited an ELAN component, a well-established marker of automatic syntactic parsing and initial structure building (e.g., Luyao Chen et al. 2023; Friederici 2004; Koelsch et al. 2005). This suggests that bilinguals may employ comparable strategies for early lexical processing across their two languages. Moreover, in L1 Chinese, morphological priming elicited stronger ELAN responses than semantic priming, whereas no such difference was detected in L2 English. Thus, early structural sensitivity appears to differ in degree between L1 and L2, rather than in kind. Notably, ELAN has often been considered an indicator of "native-likeness" in L2 morpho-syntactic acquisition (Hahne 2001; Hahne and Friederici 2001; Mueller 2005), raising the possibility that with sufficient exposure and proficiency, bilinguals could develop ELAN responses comparable to their L1.

In the 200–300 ms time window, we observed a P250 effect, with semantic priming producing stronger positivity than morphological priming over the left hemisphere. The P250 has been linked to early semantic processing under short SOAs in both German (Hill et al. 2002) and Chinese (K. K. Chung et al. 2010). In particular, as revealed by our Chap. 7, semantically related word pairs often evoke greater P250 amplitudes than morphologically related pairs, which has been taken as an index of automatic structural priming and morphological parsing. The present findings replicate and extend this pattern to Chinese derivational words, and importantly, the effect was evident in both L1 and L2. These results suggest that P250 may reflect an affix-stripping process common to both languages, providing a neural correlate of morphological priming facilitation, even though the behavioral and ELAN results revealed such facilitation only in L1 Chinese.

Following the P250, we observed a sustained negativity over left anterior sites between 300–500 ms. We refer to this effect more broadly as a late positivity/negativity complex, which may index different underlying mechanisms. On the one hand, its distribution is compatible with an LAN-like effect, typically associated with morpho-syntactic error monitoring and structural reanalysis in both sentence processing and word-level parsing (Bölte et al. 2009; Lang Chen et al. 2007; Morris and Holcomb 2005; Rossi et al. 2006). On the other hand, an alternative account is that this pattern reflects an anterior N400, a classical component linked to lexico-semantic integration (Janssen et al. 2006; Leinonen et al. 2008). Although we found evidence for morphological priming facilitation in this time window, no significant interaction between language and priming type emerged. Taken together, the late component is most plausibly interpreted as a controlled evaluation of morphological constraints, potentially reflecting both structural and semantic processes in bilingual word reading.

8.7 Conclusion

The present study combined behavioral measures, ERP, and fNIRS to investigate how adult Chinese–English bilinguals process morphology in their two languages. By using a derivational priming paradigm that minimized structural mismatches, we identified both convergent and divergent patterns of bilingual morphological processing. The fNIRS data revealed a dissociation between morphological and semantic priming in the left frontal and temporal regions: morphological parsing relied more strongly on the left prefrontal cortex in L1 than L2, while semantic association preferentially engaged the left STG and MTG. Meanwhile, ERP analyses showed that bilinguals shared common sub-lexical mechanisms across languages, such as affix-stripping and structural priming indexed by P250, yet differed in the robustness of ELAN, which was stronger in L1 Chinese than in L2 English. Together, these findings suggest that bilingual morphological processing is supported by a shared fronto-temporal network, with language-specific differences emerging in the degree of engagement and sensitivity.

These results resonate with the transfer facilitation model (Koda 2008), which posits that overlap in linguistic structures enables L1 knowledge to facilitate L2 processing. In our case, the use of derivational morphology, present in both Chinese and English, allowed bilinguals to draw on common parsing strategies, thereby transferring L1 resources into L2 word recognition. At the same time, the findings also align with the interactive framework (S. C. Chung et al. 2019a), which emphasizes that transfer is not uniform but shaped by multiple interacting factors, including typological similarity, proficiency, and task demands. Our results highlight that even when structural overlap exists, L1 and L2 may differ in the strength and automaticity of their neural correlates, reflecting the dynamic nature of cross-language transfer.

To refine this theoretical picture, future work must examine a broader range of individual variables. Proficiency, reading ability, and AoA are all likely moderators of the balance between shared and language-specific processing. In fact, we are currently conducting an fMRI study on proficiency effects in bilingual morphological processing. However, due to space and scope limitations, those results cannot be included in the present volume. Moving forward, integrating such factors will be essential for testing how bilingual morphological processing varies across learners, and for identifying the specific conditions under which L1 resources can effectively scaffold L2 development.

In sum, the present study provides converging evidence that bilingual morphological processing is best understood as an interactive, dynamic system. L1 and L2 draw on common neural resources where structures overlap, but also diverge in degree depending on language dominance and individual profiles. This dual perspective underscores the need for models of bilingual reading that integrate transfer facilitation with interactive, proficiency-dependent dynamics, thereby offering a more nuanced account of how bilinguals build and negotiate morphological representations across their two languages.

References

Bakker, I., MacGregor, L. J., Pulvermüller, F., & Shtyrov, Y. (2013). Past tense in the brain's time: Neurophysiological evidence for dual-route processing of past-tense verbs. *Neuroimage, 71*, 187-195.

Bates, D., Mächler, M., Bolker, B., & Walker, S. (2014). Fitting linear mixed-effects models using lme4. *arXiv preprint arXiv:1406.5823*.

Bick, A., Goelman, G., & Frost, R. (2008). Neural correlates of morphological processes in Hebrew. *Journal of Cognitive Neuroscience, 20*(3), 406-420.

Bölte, J., Jansma, B. M., Zilverstand, A., & Zwitserlood, P. (2009). Derivational morphology approached with event-related potentials. *The Mental Lexicon, 4*(3), 336-353.

Brysbaert, M., & New, B. (2009). Moving beyond Kučera and Francis: A critical evaluation of current word frequency norms and the introduction of a new and improved word frequency measure for American English. *Behavior research methods, 41*(4), 977-990.

Cai, Q., & Brysbaert, M. (2010). SUBTLEX-CH: Chinese word and character frequencies based on film subtitles. *PLoS One, 5*(6), e10729.

Chen, L., Shu, H., Liu, Y., Zhao, J., & Li, P. (2007). ERP signatures of subject–verb agreement in L2 learning. *Bilingualism: Language and Cognition, 10*(2), 161-174.

Chen, L., Yang, M., Gao, F., Fang, Z., Wang, P., & Feng, L. (2023). Mandarin Chinese L1 and L2 complex sentence reading reveals a consistent electrophysiological pattern of highly interactive syntactic and semantic processing: An ERP study. *Frontiers in psychology, 14*, 1143062.

Cheng, X., Yin, L., & Zhang, H. (2025). Morphological awareness and vocabulary knowledge in monolingual and bilingual learners: A meta-analysis. *International Journal of Bilingualism*, 13670069241311029.

Chung, K. K., Tong, X., Liu, P. D., McBride-Chang, C., & Meng, X. (2010). The processing of morphological structure information in Chinese coordinative compounds: an event-related potential study. *Brain Res, 1352*, 157-166. doi:https://doi.org/10.1016/j.brainres.2010.06.069

Chung, S. C., Chen, X., & Geva, E. (2019a). Deconstructing and reconstructing cross-language transfer in bilingual reading development: An interactive framework. *Journal of Neurolinguistics, 50*, 149-161.

Chung, W., Park, M.-K., & Kim, S. Y. (2019b). An electrophysiological investigation of translation and morphological priming in biscriptal bilinguals. *Journal of Neurolinguistics, 51*, 151-164. https://doi.org/10.1016/j.jneuroling.2019.01.002

Clahsen, H., & Felser, C. (2006). Grammatical processing in language learners. *Applied Psycholinguistics, 27*(1), 3-42.

Clahsen, H., & Felser, C. (2018). Some notes on the shallow structure hypothesis. *Studies in Second Language Acquisition, 40*(3), 693-706.

Forgács, B., Bohrn, I., Baudewig, J., Hofmann, M. J., Pléh, C., & Jacobs, A. M. (2012). Neural correlates of combinatorial semantic processing of literal and figurative noun noun compound words. *Neuroimage, 63*(3), 1432-1442.

Friederici, A. D. (2004). Event-related brain potential studies in language. *Current neurology and neuroscience reports, 4*(6), 466-470.

Fruchter, J., Stockall, L., & Marantz, A. (2013). MEG masked priming evidence for form-based decomposition of irregular verbs. *Frontiers in human neuroscience, 7*, 798.

Gu, J., Yu, L., & Ma, P. (2012). P2a reflects morphology structural processes in brain: Evidence from ERPs [in Chinese]. *Journal of Southwest University (Natural Science Edition), 34*(2), 137-144.

Hahne, A. (2001). What's different in second-language processing? Evidence from event-related brain potentials. *Journal of psycholinguistic research, 30*(3), 251-266.

Hahne, A., & Friederici, A. D. (2001). Processing a second language: Late learners' comprehension mechanisms as revealed by event-related brain potentials. *Bilingualism: Language and Cognition, 4*(2), 123-141.

Hao, Y., Wu, C., & Duan, X. (2024). Processing mandarin Chinese compound words by native speakers and second language learners: word frequency, semantic transparency, and word structure. *Sage Open, 14*(2), 21582440241256249.

Hill, H., Strube, M., Roesch-Ely, D., & Weisbrod, M. (2002). Automatic vs. controlled processes in semantic priming—differentiation by event-related potentials. *International Journal of Psychophysiology, 44*(3), 197-218.

Hsu, C.-H., Pylkkänen, L., & Lee, C.-Y. (2019). Effects of morphological complexity in left temporal cortex: An MEG study of reading Chinese disyllabic words. *Journal of Neurolinguistics, 49*, 168-177. https://doi.org/10.1016/j.jneuroling.2018.06.004

Ip, K. I., Hsu, L. S., Arredondo, M. M., Tardif, T., & Kovelman, I. (2017). Brain bases of morphological processing in Chinese-English bilingual children. *Dev Sci, 20*(5). https://doi.org/10.1111/desc.12449

Ip, K. I., Marks, R. A., Hsu, L. S., Desai, N., Kuan, J. L., Tardif, T., & Kovelman, L. (2019). Morphological processing in Chinese engages left temporal regions. *Brain Lang, 199*, 104696. https://doi.org/10.1016/j.bandl.2019.104696

Janssen, U., Wiese, R., & Schlesewsky, M. (2006). Electrophysiological responses to violations of morphosyntactic and prosodic features in derived German nouns. *Journal of Neurolinguistics, 19*(6), 466-482.

Ke, S., Miller, R. T., Zhang, D., & Koda, K. (2021). Crosslinguistic sharing of morphological awareness in biliteracy development: A systematic review and meta-analysis of correlation coefficients. *Language Learning, 71*(1), 8-54.

Koda, K. (2008). Impacts of prior literacy experience on second language learning to read. In *Learning to read across languages* (pp. 80-108): Routledge.

Koelsch, S., Gunter, T. C., Wittfoth, M., & Sammler, D. (2005). Interaction between syntax processing in language and in music: An ERP study. *Journal of Cognitive Neuroscience, 17*(10), 1565-1577.

Kuznetsova, A., Brockhoff, P. B., & Christensen, R. H. B. (2015). Package 'lmertest'. *R package version, 2*(0), 734.

Lehtonen, M., Vorobyev, V., Soveri, A., Hugdahl, K., Tuokkola, T., & Laine, M. (2009). Language-specific activations in the brain: Evidence from inflectional processing in bilinguals. *Journal of Neurolinguistics, 22*(5), 495-513.

Lehtonen, M., Vorobyev, V. A., Hugdahl, K., Tuokkola, T., & Laine, M. (2006). Neural correlates of morphological decomposition in a morphologically rich language: An fMRI study. *Brain and Language, 98*(2), 182-193.

Leinonen, A., Brattico, P., Jarvenpaa, M., & Krause, C. M. (2008). Event-related potential (ERP) responses to violations of inflectional and derivational rules of Finnish. *Brain Res*, 1218, 181-193.

Lemhöfer, K., & Broersma, M. (2012). Introducing LexTALE: A quick and valid lexical test for advanced learners of English. *Behavior research methods, 44*(2), 325-343.

Leminen, A., Smolka, E., Dunabeitia, J. A., & Pliatsikas, C. (2019). Morphological processing in the brain: The good (inflection), the bad (derivation) and the ugly (compounding). *Cortex, 116*, 4-44. https://doi.org/10.1016/j.cortex.2018.08.016

Li, B., Ding, Z., De Deyne, S., & Cai, Q. (2024). A large-scale database of Mandarin Chinese word associations from the Small World of Words Project. *Behav Res Methods, 57*(1), 34. https://doi.org/10.3758/s13428-024-02513-1

Liu, L., Tao, R., Wang, W., You, W., Peng, D., & Booth, J. R. (2013). Chinese dyslexics show neural differences in morphological processing. *Dev Cogn Neurosci, 6*, 40-50. https://doi.org/10.1016/j.dcn.2013.06.004

MacWhinney, B., & Kroll, J. F. (2005). A unified model of language acquisition. *Handbook of bilingualism: Psycholinguistic approaches, 4967*, 50-70.

Marks, R. A., Labotka, D., Sun, X., Nickerson, N., Eggleston, R. L., Yu, C.-L., ... Kovelman, I. (2021). Morphological awareness contributes to early literacy in linguistically diverse readers. *PsyArXiv*

Morris, J., & Holcomb, P. J. (2005). Event-related potentials to violations of inflectional verb morphology in English. *Cognitive Brain Research, 25*(3), 963-981.

Morris, J., & Stockall, L. (2012). Early, equivalent ERP masked priming effects for regular and irregular morphology. *Brain and Language, 123*(2), 81-93.

Mueller, J. L. (2005). Electrophysiological correlates of second language processing. *Second Language Research, 21*(2), 152-174.

Nelson, D. L., McEvoy, C. L., & Schreiber, T. A. (2004). The University of South Florida free association, rhyme, and word fragment norms. *Behavior Research Methods, Instruments, & Computers, 36*(3), 402-407.

Prehn, K., Taud, B., Reifegerste, J., Clahsen, H., & Flöel, A. (2018). Neural correlates of grammatical inflection in older native and second-language speakers. *Bilingualism: Language and Cognition, 21*(1), 1-12.

Rossi, S., Gugler, M. F., Friederici, A. D., & Hahne, A. (2006). The impact of proficiency on syntactic second-language processing of German and Italian: Evidence from event-related potentials. *Journal of Cognitive Neuroscience, 18*(12), 2030-2048.

Schiller, N. O. (2020). Neurolinguistic Approaches in Morphology. In *Oxford Research Encyclopedia of Linguistics*.

Sun, J., Luo, X., & Pae, H. K. (2024). The Recognition of Chinese Compound Words by Native English-and Korean-speaking Learners of Chinese. *Reading in a foreign language, 36*(1), 1-22.

Sun, X., Marks, R. A., Zhang, K., Yu, C. L., Eggleston, R. L., Nickerson, N., … Satterfield, T. (2023). Brain bases of English morphological processing: A comparison between Chinese-English, Spanish-English bilingual, and English monolingual children. *Developmental science, 26*(1), e13251.

Sun, X., Zhang, K., Marks, R. A., Nickerson, N., Eggleston, R. L., Yu, C. L., … Kovelman, I. (2022). What's in a word? Cross-linguistic influences on Spanish–English and Chinese–English bilingual children's word reading development. *Child development, 93*(1), 84-100.

Tan, L. H., Laird, A. R., Li, K., & Fox, P. T. (2005). Neuroanatomical correlates of phonological processing of Chinese characters and alphabetic words: a meta-analysis. *Hum Brain Mapp, 25*(1), 83-91. https://doi.org/10.1002/hbm.20134

Tan, L. H., Spinks, J. A., Feng, C. M., Siok, W. T., Perfetti, C. A., Xiong, J., … Gao, J. H. (2003). Neural systems of second language reading are shaped by native language. *Human brain mapping, 18*(3), 158-166.

Vingerhoets, G., Van Borsel, J., Tesink, C., van Den Noort, M., Deblaere, K., Seurinck, R., … Achten, E. (2003). Multilingualism: an fMRI study. *Neuroimage, 20*(4), 2181-2196.

Wattendorf, E., Westermann, B., Zappatore, D., Franceschini, R., Lüdi, G., Radü, E.-W., & Nitsch, C. (2001). Different languages activate different subfields in Broca area. *Neuroimage, 6*(13), 624.

Whiting, C. M., Marslen-Wilson, W. D., & Shtyrov, Y. (2013). Neural dynamics of inflectional and derivational processing in spoken word comprehension: laterality and automaticity. *Frontiers in human neuroscience, 7*, 759.

Xu, Y., Graber, H. L., Schmitz, C. H., & Barbour, R. L. (2014). *nirs LAB: A Problem-Solving Environment for fNIRS Neuroimaging Data Analysis*.

Yang, X. (2020). *A semantic logic analysis on Chinese quasi-affixes and its pedagogical implications* [in Chinese] (Master's thesis). Liaoning Normal University.

Ye, J. C., Tak, S., Jang, K. E., Jung, J., & Jang, J. (2009). NIRS-SPM: statistical parametric mapping for near-infrared spectroscopy. *Neuroimage, 44*(2), 428-447.

Zhao, S., Wu, Y., Tsang, Y.-K., Sui, X., & Zhu, Z. (2021). Morpho-semantic analysis of ambiguous morphemes in Chinese compound word recognition: An fMRI study. *Neuropsychologia, 157*, 107862.

Zou, L., Packard, J. L., Xia, Z., Liu, Y., & Shu, H. (2015). Neural Correlates of Morphological Processing: Evidence from Chinese. *Front Hum Neurosci, 9*, 714. https://doi.org/10.3389/fnhum.2015.00714

Zou, L., Packard, J. L., Xia, Z., Liu, Y., & Shu, H. (2019). Morphological and Whole-Word Semantic Processing Are Distinct: Event Related Potentials Evidence From Spoken Word Recognition in Chinese. *Front Hum Neurosci, 13*, 133. https://doi.org/10.3389/fnhum.2019.00133

Chapter 9
General Discussion and Conclusion

9.1 Summary of Main Findings

This book has explored the cognitive and neural mechanisms of Chinese morphological processing, integrating evidence from corpus analyses, behavioral paradigms, and multimodal neuroimaging techniques (EEG/ERP, fNIRS). The central aim has been to clarify the representational units, temporal dynamics, and spatial localization of morphology in Chinese word recognition, and to extend these findings to the bilingual context. Across Chaps. 5–8, the results consistently supported a dual-route account in which whole-word and morphemic representations coexist, but their relative contributions depend on linguistic properties (e.g., morpheme boundness, structural legality) as well as individual learner factors (e.g., proficiency, L1 background).

A key theoretical clarification of this book is the working definition of Chinese morphology. Rather than focusing on inflectional paradigms typical of Indo-European languages, this book adopts "构词法" (*word formation*) as its central lens: the study of how morphemes combine through compounding, derivation, reduplication, or other processes to form new lexical items. This perspective emphasizes both the external productivity of word formation and the internal organization of word structure, aligning more closely with the typological properties of Chinese. Importantly, this definition also enables cross-linguistic comparison by situating Chinese within the broader landscape of morphology while foregrounding its unique characteristics.

Taken together, the evidence presented in this book offers a comprehensive account of when morphological information is accessed during Chinese word reading, where in the brain such processes are localized, and how morphological representations are integrated with semantic processing and lexical development. In the following sections, we summarize these findings with respect to the three guiding questions of the when, the where, and the how of Chinese morphological processing.

F. Gao, *The When, Where, and How of Morphological Processing in the Human Brain*, https://doi.org/10.1007/978-981-95-5981-7_9

9.1.1 The When

High-temporal-resolution ERP evidence revealed that Chinese morphological processing and semantic processing can be distinguished in terms of their time course. As shown in Chap. 6, morphological encoding tends to engage relatively later controlled and conscious resources, resembling syntactic processing, whereas semantic processing emerges earlier. This temporal distinction highlights the unique status of Chinese morphology, which recruits cognitive resources more akin to sentence-level parsing than to automatic lexical semantics.

At the same time, Chaps. 7 and 8 identified early ERP components (notably P250 and ELAN) associated with morphological structure. We suggest that these effects arise from the specific requirements of the priming paradigms and the attentional processes they engage. Crucially, ELAN was observed across both L1 and L2 processing, marking automatic structural parsing in priming conditions. In L1 Chinese, morphological priming elicited stronger ELAN responses than semantic priming, whereas in L2 English this distinction was less significant. Thus, early sensitivity to morphological structure appears to differ in degree between L1 and L2 rather than in kind, suggesting comparable strategies but different robustness across languages.

9.1.2 The Where

Converging fNIRS evidence across Chaps. 6–7 consistently localized Chinese morphological processing to the left frontal cortex. This region appears to serve as the neural hub for evaluating structural legality, coordinating morphemic integration, and managing morpho-semantic constraints. In Chinese L1 contexts, the left frontal cortex thus provides the most stable marker of morphological encoding.

In bilingual contexts (Chap. 8), activity in the left prefrontal cortex also emerged as the most robust indicator of shared morphological processing across L1 Chinese and L2 English, highlighting its centrality to cross-linguistic word-formation mechanisms. This finding underscores the notion that despite typological differences, bilinguals recruit a common frontal network for morphological parsing, albeit with language-specific differences in robustness and automaticity.

Although no strong left temporal activations were consistently observed in native Chinese visual word recognition, prior auditory studies have implicated the temporal cortex in morphological processing (e.g., Ip et al. 2019). Importantly, Chap. 7 revealed significant correlations between P250 and HbO responses in both frontal and temporal regions, especially in the derivation–compound contrast, suggesting shared dynamics across electrophysiological and hemodynamic measures. Moreover, Chap. 8 demonstrated that the left temporal cortex (MTG and STG) remains important for differentiating semantic associations from morphological relationships, often showing stronger engagement for semantic processing.

Thus, while the left frontal cortex dominates morphological processing, the temporal cortex may still play a supporting role, particularly in semantic–morphological interactions. Given the spatial limitations of fNIRS, future work should employ fMRI and MEG to further refine our understanding of where morphology is processed in the brain.

9.1.3 The How

Both behavioral and neuroimaging evidence contribute to addressing the question of *how* Chinese morphological processing unfolds. Behavioral findings (Chap. 5) demonstrated that lexical representation in both L1 and L2 Chinese supports a dual-route model. While native speakers preferentially relied on economical whole-word representations, L2 learners showed a developmental continuum from morpheme-based to whole-word reliance, with free morphemes in particular facilitating decomposition more strongly than bound morphemes. Neuroimaging evidence (Chaps. 6–8) further revealed that these dual routes are instantiated in distinct spatio-temporal patterns. EEG results showed that morphology engages later controlled processes (e.g., late N400) as well as earlier structure-sensitive components under priming conditions (e.g., P250, ELAN), whereas fNIRS localized morphological parsing to the left prefrontal cortex, with additional modulation from temporal regions in derivation–compound contrasts and in bilingual contexts. Together, these findings highlight how linguistic factors such as morpheme type and structural legality, alongside individual factors such as proficiency and bilingual experience, jointly shape the balance between whole-word and morphemic routes in Chinese word representation and processing. Notably, compared with alphabetic languages where morphology is often tied to inflectional paradigms and affixation, Chinese morphological processing is largely realized through compounding and reflects more controlled operations, underscoring the language-specific pathways through which the dual-route architecture manifests.

Neuroimaging evidence revealed a functional dissociation between morphological and semantic processing. In Indo-European languages, morphology is typically situated at the *lexeme* level of the mental lexicon (Levelt 1989), reflecting inflectional alternations across word forms (e.g., run–runs–running). By contrast, our findings suggest that Chinese morphology is more appropriately represented at the *lemma* level, where higher-order structural and combinatory information is encoded rather than surface word forms. This reconceptualization aligns with two key typological features of Chinese. First, Chinese morphology relies predominantly on compounding rather than inflection, with new words formed through the systematic combination of morphemes into disyllabic or polysyllabic units. Second, morphological processing in Chinese often engages controlled, higher-level combinatory mechanisms that closely resemble sentence-like parsing operations in their neural profile. Together, these results indicate that Chinese morphology occupies a distinctive representational locus within the mental lexicon, one that bridges word formation and syntactic computation, thereby highlighting both language-specific adaptations and broader cross-linguistic principles of morphological processing.

Additionally, in bilingual contexts, morphological processing demonstrates both consistencies and divergences across languages, whether the transfer is from another language into Chinese or from Chinese into another language such as English. In Chap. 5, we showed that Chinese L2 learners exhibit a developmental continuum in their lexical representations, shifting from morpheme-based to whole-word reliance. This pattern suggests that learners actively adapt their processing strategies rather than simply transferring structural templates from their L1. In other words, our emphasis is less on structural transfer, yet more on the transformation of processing strategies, which are dynamically reconfigured as proficiency grows. In Chap. 8, by focusing on morphological structures shared between Chinese and English, we tested whether bilinguals benefit from transfer facilitation (Koda 2008). Results revealed evidence for both convergence (e.g., shared sub-lexical mechanisms such as affix-stripping) and divergence, with Chinese morphology eliciting stronger and more automatic neural responses than English. Collectively, these findings underscore the dynamic and interactive nature of bilingual morphological processing (Chung et al. 2019). Transfer is neither uniform nor unidirectional, but shaped by multiple interacting factors, including typological similarity between the languages, learners' proficiency levels, and task demands. This perspective highlights bilingualism as a flexible system in which cognitive strategies are continuously negotiated across languages rather than rigidly copied from one to the other, while also shedding light on the universal versus language-specific mechanisms that underlie morphological processing in the human brain.

In sum, the combined evidence across Chaps. 5–8 clarifies when, where, and how Chinese morphological processing occurs in the brain. By adopting a word-formation perspective on morphology, this book contributes not only to Chinese psycholinguistics but also to cross-linguistic theories of morphology, offering new insights into the architecture of the mental lexicon and the neurocognitive principles that support language learning and bilingual processing.

9.2 Implications and Future Directions

9.2.1 Methodological Insights

We advocate a three-stage research framework for the study of morphological processing, and reading research more broadly: **corpus-driven observation, experimental investigation, and implications/applications**. First, the identification of research questions should be grounded in real-world language use, drawing on both native speakers' and learners' corpora to reveal factors that influence processing, pose difficulties, or generate systematic errors. Corpus-based analyses not only highlight issues of practical and theoretical significance, such as lexical errors in

Chinese L2 production that motivated our focus on representational units, but also provide valuable resources for selecting ecologically valid experimental materials. This illustrates the complementary and indispensable role of corpus-driven approaches and experimental research.

In the experimental stage, **offline and online methods should be viewed as mutually reinforcing**. For instance, studies of morphological awareness typically rely on offline measures (e.g., questionnaires, interviews), whereas investigations of morphological processing often employ online time-pressured methods. Behavioral paradigms yield indirect indicators such as reaction times, while neuroimaging techniques provide direct physiological measures of brain activity. Importantly, offline and online approaches, as well as behavioral and neuroimaging data, are mutually complementary, together offering a more comprehensive depiction of the mechanisms underlying morphological processing. A key innovation of this book lies in the combined use of EEG and fNIRS, which enables fine-grained tracking of both the temporal dynamics and spatial localization of Chinese morphological processing. This multimodal approach supplies crucial neural evidence for the dual-route model and establishes methodological references for future reading research. Importantly, the findings can then cycle back to the real world, informing pedagogical practices and clinical interventions.

Meanwhile, we highlight the importance of adopting a **cross-linguistic perspective** that integrates linguistic typology and learning sciences. Research on Chinese not only supports, modifies, or extends models originally based on Indo-European languages, but also deepens our understanding of the interplay between language-specific and universal mechanisms of morphology (Perfetti et al. 2013; Share 2008). Bilingual investigations further demonstrate the generalizability of methods, while showcasing the value of typological diversity and providing insights for language learning science.

Finally, this book underscores the necessity of **open science practices**, including data sharing, megastudies, and the construction of large-scale corpora and neuroimaging databases. For Chinese morphology in particular, building standardized normative resources such as frequency norms, semantic transparency ratings, and morphological structure annotations will be essential for reproducibility and cross-study comparability. Beyond database construction, the development of user-friendly research tools and computational platforms will greatly facilitate the accessibility and application of these resources across disciplines. Such efforts will enable researchers to probe structural differences in morphology and its interactions with other linguistic domains such as semantics and syntax, thereby advancing our understanding of how morphology constrains and enables broader language structure in the human brain (e.g., the hierarchical nature of language, Ding et al. 2016; Sheng et al. 2019). We argue that these directions define the next decade of Chinese morphological processing research.

At the same time, it is important to acknowledge a limitation of the present book. All experimental paradigms we employed focused on the processing of isolated words, which inevitably constrains ecological validity and cannot fully capture the dynamics of real reading. Future research should therefore extend these investigations to sentence-level reading and naturalistic contexts, exploring how morphological processing unfolds in more complex environments and how its neural basis interacts with higher-order discourse and comprehension processes.

9.2.2 Pedagogical and Clinical Implications

At the pedagogical level, this book underscores that morphology is a core component of language development, in both first and second languages, and that its neural representation can be distinguished from other linguistic components. Morphological knowledge involves not only implicit morphological processing but also explicit morphological awareness. Drawing on findings from Chap. 5, we argue for level-sensitive vocabulary instruction: beginners should prioritize whole-word learning with limited morpheme exposure, intermediate learners benefit from an emphasis on free morphemes as scaffolds for lexical expansion, and advanced learners may adopt whole-word strategies while still maintaining sensitivity to morphemic structure. Such a staged approach avoids the pitfalls of "one-size-fits-all" pedagogy. In addition, L2 instruction should account for both cross-linguistic similarities and differences between learners' L1 and L2, while also recognizing the transfer of processing strategies across languages (Chap. 8).

Another pedagogical implication concerns the value of providing learners with appropriate metalinguistic knowledge of morphology. Explicit instruction in morphological concepts can enhance morphological awareness, which in turn supports vocabulary acquisition and reading development. Nevertheless, it is important to acknowledge a limitation of this book: we did not conduct classroom-based intervention studies. Future research should adopt an educational neuroscience perspective (Mareschal et al. 2013; Thomas et al. 2019), linking classroom instruction with brain-based measures of morphological processing. Such studies would help bridge the gap between morphological awareness and morphological processing, yielding more directly applicable recommendations for L2 teaching.

At the clinical level, research on morphological processing has implications for the diagnosis and rehabilitation of language disorders, including aphasia and dyslexia. Because morphological operations interact closely with semantic and syntactic processing, understanding these mechanisms can inform the design of individualized treatment plans. The neuroimaging evidence presented in this book, particularly regarding the role of the left frontal cortex, may contribute to more precise neuro-linguistic localization and the development of targeted neuro-intervention protocols.

Finally, the findings have broader cross-cultural and cross-linguistic applications. Insights from Chinese morphological processing can inform multilingual education, contribute to language policy (e.g., curriculum design that emphasizes morphology in literacy development), and inspire advances in artificial intelligence, particularly in natural language processing systems that require sensitivity to morpheme-based structures. By situating Chinese within a comparative perspective, this book demonstrates that research on morphology not only advances psycholinguistic theory but also carries translational value for education, clinical practice, and information technology. Looking forward, progress in this area will rely increasingly on interdisciplinary collaboration across linguistics, neuroscience, education, and computational sciences, ensuring that methodological advances are translated into both theoretical innovation and practical impact.

References

Chung, S. C., Chen, X., & Geva, E. (2019). Deconstructing and reconstructing cross-language transfer in bilingual reading development: An interactive framework. *Journal of Neurolinguistics, 50*, 149-161.

Ding, N., Melloni, L., Zhang, H., Tian, X., & Poeppel, D. (2016). Cortical tracking of hierarchical linguistic structures in connected speech. *Nat Neurosci, 19*(1), 158-164. doi:https://doi.org/10.1038/nn.4186

Ip, K. I., Marks, R. A., Hsu, L. S., Desai, N., Kuan, J. L., Tardif, T., & Kovelman, L. (2019). Morphological processing in Chinese engages left temporal regions. *Brain Lang, 199*, 104696. doi:https://doi.org/10.1016/j.bandl.2019.104696

Koda, K. (2008). Impacts of prior literacy experience on second language learning to read. In *Learning to read across languages* (pp. 80-108). Routledge.

Levelt, W. J. (1989). *Speaking: From intention to articulation.* MIT Press.

Mareschal, D., Butterworth, B., & Tolmie, A. (2013). *Educational neuroscience.* Wiley Online Library.

Perfetti, C., Cao, F., & Booth, J. (2013). Specialization and Universals in the Development of Reading Skill: How Chinese Research Informs a Universal Science of Reading. *Sci Stud Read, 17*(1), 5-21. doi:https://doi.org/10.1080/10888438.2012.689786

Share, D. L. (2008). On the Anglocentricities of current reading research and practice: the perils of overreliance on an "outlier" orthography. *Psychol Bull, 134*(4), 584-615. doi:https://doi.org/10.1037/0033-2909.134.4.584

Sheng, J., Zheng, L., Lyu, B., Cen, Z., Qin, L., Tan, L. H., ... Gao, J.-H. (2019). The cortical maps of hierarchical linguistic structures during speech perception. *Cerebral Cortex, 29*(8), 3232-3240.

Thomas, M. S., Ansari, D., & Knowland, V. C. (2019). Annual research review: Educational neuroscience: Progress and prospects. *Journal of child psychology and psychiatry, 60*(4), 477-492.

Appendix A: Full List of the Experimental Stimuli of Chap. 5

Morpheme Boundness	Morpheme			Whole-word		
	Stimuli	Pinyin	Meaning	Stimuli	Pinyin	Meaning
Free	爱	ài	love	爱人	ài rén	spouse
Free	北	běi	north	北边	běi biān	north side
Free	变	biàn	change	变化	biàn huà	change
Free	车	chē	vehicle	车站	chē zhàn	bus station
Free	城	chéng	city	城市	chéng shì	city
Free	电	diàn	electricity	电灯	diàn dēng	electric light
Free	饭	fàn	meal	饭店	fàn diàn	restaurant
Free	国	guó	country	国家	guó jiā	country
Free	黑	hēi	black	黑板	hēi bǎn	blackboard
Free	课	kè	lesson	课本	kè běn	textbook
Free	跑	pǎo	run	跑步	pǎo bù	run
Free	声	shēng	voice	声音	shēng yīn	voice
Free	睡	shuì	sleep	睡觉	shuì jiào	sleep
Free	说	shuō	speak	说明	shuō míng	explain
Free	天	tiān	sky	天气	tiān qì	weather
Free	晚	wǎn	late	晚会	wǎn huì	evening party
Free	新	xīn	new	新年	xīn nián	New Year
Free	学	xué	study	学习	xué xí	study
Free	月	yuè	moon	月亮	yuè liang	moon
Free	早	zǎo	early	早晨	zǎo chén	morning
Free	真	zhēn	true	真正	zhēn zhèng	genuine
Free	中	zhōng	middle	中间	zhōng jiān	middle
Free	帮	bāng	help	帮助	bāng zhù	assistance
Free	困	kùn	trapped	困难	kùn nán	difficulty
Free	眼	yǎn	eye	眼睛	yǎn jing	eye
Free	进	jìn	enter	进去	jìn qù	enter

F. Gao, *The When, Where, and How of Morphological Processing in the Human Brain*, https://doi.org/10.1007/978-981-95-5981-7

Morpheme Boundness	Morpheme			Whole-word		
	Stimuli	Pinyin	Meaning	Stimuli	Pinyin	Meaning
Free	牛	niú	cow	牛奶	niú nǎi	milk
Free	错	cuò	wrong	错误	cuò wù	mistake
Free	见	jiàn	see	见面	jiàn miàn	meet
Free	比	bǐ	compare	比较	bǐ jiào	compare
Bound	产	chǎn	produce	产品	chǎn pǐn	product
Bound	参	cān	participate	参加	cān jiā	participate
Bound	访	fǎng	visit	访问	fǎng wèn	visit
Bound	丰	fēng	abundant	丰富	fēng fù	abundant
Bound	汉	hàn	Han (ethnic group)	汉字	hàn zì	Chinese character
Bound	欢	huān	joyous	欢迎	huān yíng	welcome
Bound	民	mín	people	民族	mín zú	ethnic group
Bound	朋	péng	friend	朋友	péng you	friend
Bound	特	tè	special	特别	tè bié	special
Bound	危	wēi	danger	危险	wēi xiǎn	danger
Bound	文	wén	culture	文化	wén huà	culture
Bound	希	xī	hope	希望	xī wàng	hope
Bound	喜	xǐ	happy	喜欢	xǐ huān	like
Bound	研	yán	research	研究	yán jiū	research
Bound	衣	yī	clothes	衣服	yī fu	clothes
Bound	游	yóu	swim	游泳	yóu yǒng	swim
Bound	政	zhèng	politics	政府	zhèng fǔ	government
Bound	知	zhī	to know	知道	zhī dào	know
Bound	注	zhù	note	注意	zhù yì	pay attention
Bound	安	ān	safe	安全	ān quán	security
Bound	保	bǎo	guarantee	保证	bǎo zhèng	guarantee
Bound	语	yǔ	language	语法	yǔ fǎ	grammar
Bound	永	yǒng	eternal	永远	yǒng yuǎn	forever
Bound	今	jīn	today	今天	jīn tiān	today
Bound	机	jī	opportunity	机会	jī huì	opportunity
Bound	医	yī	medicine	医生	yī shēng	doctor
Bound	意	yì	idea	意思	yì si	meaning
Bound	主	zhǔ	master	主动	zhǔ dòng	initiative
Bound	足	zú	foot	足球	zú qiú	football
Bound	厕	cè	toilet	厕所	cè suǒ	toilet

Appendix B: Full List of the Experimental Stimuli of Chap. 6

Category	No.	Stimuli	Pinyin	Word meaning	Literal meaning	Structure
Real word	1	丰富	fēng fù	abundant	abundant-rich	Coordinate
Real word	2	危险	wēi xiǎn	dangerous	dangerous-dangerous	Coordinate
Real word	3	简单	jiǎn dān	simple	simple-simple	Coordinate
Real word	4	并联	bìng lián	parallel connection	parallel-connection	Subordinate
Real word	5	北边	běi bian	north	north-side	Subordinate
Real word	6	城市	chéng shì	city	town-city	Coordinate
Real word	7	饭店	fàn diàn	restaurant	meal-shop	Subordinate
Real word	8	炎症	yán zhèng	inflammation	inflammation-symptom	Coordinate
Real word	9	汉字	hàn zì	Chinese character	Chinese-character	Subordinate
Real word	10	黑板	hēi bǎn	blackboard	black-board	Subordinate
Real word	11	感官	gǎn guān	sense organ	sense-organ	Subordinate
Real word	12	农民	nóng mín	farmer	agricultural-people	Subordinate
Real word	13	阳光	yáng guāng	sunlight	sun-light	Subordinate
Real word	14	衣服	yī fú	clothes	clothes-clothes	Coordinate
Real word	15	医生	yī shēng	doctor	medical-people	Subordinate
Real word	16	肃清	sù qīng	root out	clean-clear	Verb–Resultative
Real word	17	说明	shuō míng	explain	speak-bright	Verb–Resultative
Real word	18	希望	xī wàng	hope	hope-watch	Coordinate
Real word	19	喜欢	xǐ huān	like	like-like	Coordinate
Real word	20	游泳	yóu yǒng	swim	swim-swim	Coordinate

F. Gao, *The When, Where, and How of Morphological Processing in the Human Brain*, https://doi.org/10.1007/978-981-95-5981-7

Category	No.	Stimuli	Pinyin	Word meaning	Literal meaning	Structure
Real word	21	末端	mò duān	tail end	end-side	Subordinate
Real word	22	麻烦	má fán	trouble	messy-trouble	Subordinate
Real word	23	晚安	wǎn ān	good night	night-safe	Subject-predicate
Real word	24	忘记	wàng jì	forget	forget-remember	Coordinate
Real word	25	突然	tū rán	suddenly	sudden-situation	Subordinate
Real word	26	胚胎	pēi tāi	embryo	embryo-embryo	Coordinate
Real word	27	话筒	huà tǒng	microphone	speaking-tube	Subordinate
Real word	28	季节	jì jié	season	season-festival	Coordinate
Real word	29	长城	cháng chéng	the Great Wall	long-city	Subordinate
Real word	30	债务	zhài wù	debt	debt-affair	Subordinate
Real word	31	甲板	jiǎ bǎn	deck	deck-board	Subordinate
Real word	32	活动	huó dòng	activity	things-activity	Coordinate
Real word	33	枝叶	zhī yè	branches and leaves	branch-leaf	Coordinate
Real word	34	兼顾	jiān gù	to balance	both-take care	Subordinate
Real word	35	工作	gōng zuò	work	work-do	Coordinate
Real word	36	祖父	zǔ fù	grandfather	ancestor-father	Subordinate
Real word	37	春风	chūn fēng	spring breeze	spring-wind	Subordinate
Real word	38	航海	háng hǎi	navigation	navigate-sea	Verb-Object
Real word	39	稻谷	dào gǔ	rice	rice-grain	Coordinate
Real word	40	水文	shuǐ wén	hydrology	water-situation	Subordinate
Real word	41	观赏	guān shǎng	appreciate	watch-appreciate	Coordinate
Real word	42	心理	xīn lǐ	psychology	heart-principle	Subordinate
Real word	43	罚款	fá kuǎn	penalty	fine-money	Verb-Object
Real word	44	投机	tóu jī	be opportunistic	take-chance	Verb-Object
Real word	45	扫地	sǎo dì	sweep the floor	sweep-floor	Verb-Object
Pseudoword	1	睡房	shuì fáng	/	sleep-room	Subordinate
Pseudoword	2	商牌	shāng pái	/	commercial-brand	Subordinate
Pseudoword	3	来返	lái fǎn	/	come-go back	Coordinate
Pseudoword	4	花地	huā dì	/	flower-field	Subordinate
Pseudoword	5	这刻	zhè kè	/	this-moment	Subordinate
Pseudoword	6	阅报	yuè bào	/	read-newspaper	Verb-Object
Pseudoword	7	唱者	chàng zhě	/	singing-person	Subordinate
Pseudoword	8	昨年	zuó nián	/	last-year	Subordinate
Pseudoword	9	获到	huò dào	/	harvest-obtain	Coordinate
Pseudoword	10	加电	jiā diàn	/	charge-electricity	Verb-Object
Pseudoword	11	费气	fèi qì	/	waste-gas	Verb-Object
Pseudoword	12	优处	yōu chù	/	good-point	Subordinate

Category	No.	Stimuli	Pinyin	Word meaning	Literal meaning	Structure
Pseudoword	13	半晚	bàn wǎn	/	half-night	Subordinate
Pseudoword	14	微哭	wēi kū	/	mild-crying	Subordinate
Pseudoword	15	寒期	hán qī	/	cold-period	Subordinate
Pseudoword	16	抽钱	chōu qián	/	draw-money	Verb-Object
Pseudoword	17	学具	xué jù	/	learning-tool	Subordinate
Pseudoword	18	离学	lí xué	/	leave-study	Verb-Object
Pseudoword	19	解烦	jiě fán	/	solve-trouble	Verb-Object
Pseudoword	20	兵人	bīng rén	/	soldier-man	Subordinate
Pseudoword	21	冷情	lěng qíng	/	cold-feeling	Subordinate
Pseudoword	22	夜天	yè tiān	/	night-sky	Subordinate
Pseudoword	23	看面	kàn miàn	/	look-face	Verb-Object
Pseudoword	24	外游	wài yóu	/	out-travelling	Subordinate
Pseudoword	25	岁年	suì nián	/	year-year	Coordinate
Pseudoword	26	特式	tè shì	/	special-format	Subordinate
Pseudoword	27	街路	jiē lù	/	street-road	Coordinate
Pseudoword	28	好朋	hǎo péng	/	good-friend	Subordinate
Pseudoword	29	古久	gǔ jiǔ	/	old-long ago	Coordinate
Pseudoword	30	教房	jiào fáng	/	teaching-house	Subordinate
Pseudoword	31	内边	nèi biān	/	internal-side	Subordinate
Pseudoword	32	教屋	jiào wū	/	teaching-room	Subordinate
Pseudoword	33	特处	tè chù	/	special-point	Subordinate
Pseudoword	34	忆得	yì dé	/	remember-obtained	Verb–resultative
Pseudoword	35	专碟	zhuān dié	/	specialized-disk	Subordinate
Pseudoword	36	闭课	bì kè	/	close-class	Verb-Object
Pseudoword	37	失眼	shī yǎn	/	lose-eye	Verb-Object
Pseudoword	38	钱产	qián chǎn	/	money-fortune	Coordinate
Pseudoword	39	优恶	yōu è	/	good-bad	Coordinate
Pseudoword	40	今代	jīn dài	/	this-dynasty	Subordinate
Pseudoword	41	远眼	yuǎn yǎn	/	far-sight	Subordinate
Pseudoword	42	下年	xià nián	/	next-year	Subordinate
Pseudoword	43	前果	qián guǒ	/	before-result	Subordinate
Pseudoword	44	爸母	bà mǔ	/	father-mother	Coordinate
Pseudoword	45	日饭	rì fàn	/	daytime-meal	Subordinate
Nonword	1	液必	yè bì	/	liquid-must	/
Nonword	2	者方	zhě fāng	/	person-square	/
Nonword	3	颗给	kē gěi	/	measure word-give	/
Nonword	4	针审	zhēn shěn	/	needle-censor	/
Nonword	5	素吴	sù wú	/	plain-Wu (family name)	/

Category	No.	Stimuli	Pinyin	Word meaning	Literal meaning	Structure
Nonword	6	战喃	zhàn nán	/	battle-murmur	/
Nonword	7	触批	chù pī	/	touch-batch	/
Nonword	8	题打	tí dǎ	/	question-hit	/
Nonword	9	丈巩	zhàng gǒng	/	length unit-enhance	/
Nonword	10	股肯	gǔ kěn	/	thigh-agree	/
Nonword	11	换且	huàn qiě	/	swap-function word	/
Nonword	12	的靠	de kào	/	of-rely	/
Nonword	13	吸连	xī lián	/	absorb-connect	/
Nonword	14	式借	shì jiè	/	style-borrow	/
Nonword	15	负各	fù gè	/	minus-every	/
Nonword	16	儿某	ér mǒu	/	son-some	/
Nonword	17	服但	fú dàn	/	clothes-but	/
Nonword	18	旱直	hàn zhí	/	drought-straight	/
Nonword	19	料思	liào sī	/	ingredient-think	/
Nonword	20	盐以	yán yǐ	/	salt-to	/
Nonword	21	乎共	hū gòng	/	auxiliary word-common	/
Nonword	22	育征	yù zhēng	/	breed-expropriation	/
Nonword	23	泽握	zé wò	/	shine-grasp	/
Nonword	24	吨禹	dūn yǔ	/	ton-Yu (a man's name)	/
Nonword	25	寸互	cùn hù	/	length unit-mutual	/
Nonword	26	术调	shù diào	/	skill-tune	/
Nonword	27	类更	lèi gēng	/	category-more	/
Nonword	28	样汶	yàng wèn	/	appearance-Wen (a place's name)	/
Nonword	29	柯淘	kē táo	/	family name-wash	/
Nonword	30	衬乔	chèn qiáo	/	lining-family name	/
Nonword	31	郊欠	jiāo qiàn	/	suburb-owe	/
Nonword	32	巾较	jīn jiào	/	cloth-compare	/
Nonword	33	匹因	pǐ yīn	/	measure word-because	/
Nonword	34	革使	gé shǐ	/	leather-make	/
Nonword	35	么本	me běn	/	auxiliary word-measure word	/
Nonword	36	化此	huà cǐ	/	change-this	/
Nonword	37	度很	dù hěn	/	degree-very	/
Nonword	38	平特	píng tè	/	flat-special	/
Nonword	39	象并	xiàng bìng	/	elephant-and	/
Nonword	40	代最	dài zuì	/	generation-most	/
Nonword	41	质什	zhì shí	/	quality-ten	/
Nonword	42	立或	lì huò	/	stand-or	/
Nonword	43	品反	pǐn fǎn	/	product-inverse	/
Nonword	44	次基	cì jī	/	time-foundation	/
Nonword	45	数入	shù rù	/	number-enter	/

Appendix C: The 3D MNI Coordinates and Associated Brain Regions of 22 Channels

Channel number	MNI coordinates			Anatomical label	Percentage of overlap
	x	y	z		
1	−37.6667	63.33333	−0.66667	10—Frontopolar area	0.96981
2	−48	51	−3	10—Frontopolar area	0.49259
				47—Inferior prefrontal gyrus	0.38889
				46—Dorsolateral prefrontal cortex	0.07037
3	−19.3333	65.33333	25.66667	10—Frontopolar area	0.96226
4	−8	64.66667	32	10—Frontopolar area	0.72059
5	−15.3333	55.33333	41.33333	9—Dorsolateral prefrontal cortex	0.8178
6	−36.6667	58	21.66667	10—Frontopolar area	0.95455
7	−47.6667	45.33333	19.66667	46—Dorsolateral prefrontal cortex	0.76604
				10—Frontopolar area	0.23396
8	−32.6667	47.66667	37.66667	9—Dorsolateral prefrontal cortex	0.88688
9	−47	32.33333	37.33333	9—Dorsolateral prefrontal cortex	0.5875
				46—Dorsolateral prefrontal cortex	0.33333
10	−0.33333	57.66667	37.66667	9—Dorsolateral prefrontal cortex	0.75591
11	−11	49.66667	49	8—Includes Frontal eye fields	0.77686
12	−57	29.33333	13.33333	45—pars triangularis Broca's area	0.65902
				46—Dorsolateral prefrontal cortex	0.34098
13	−56.6667	15.66667	30.33333	9—Dorsolateral prefrontal cortex	0.78292

F. Gao, *The When, Where, and How of Morphological Processing in the Human Brain*, https://doi.org/10.1007/978-981-95-5981-7

Channel number	MNI coordinates			Anatomical label	Percentage of overlap
	x	y	z		
14	−65.6667	−1	21	6—Pre-Motor and Supplementary Motor Cortex	0.6
15	−60.3333	10.33333	3.666667	22—Superior Temporal Gyrus	0.50502
16	−69.6667	−17.6667	2.666667	22—Superior Temporal Gyrus	0.37421
				42—Primary and Auditory Association Cortex	0.31447
				21—Middle Temporal gyrus	0.31132
17	−68	−6.66667	−16.3333	21—Middle Temporal gyrus	0.9571
18	−54.3333	1.333333	49.33333	6—Pre-Motor and Supplementary Motor Cortex	0.92913
19	−65	−15.3333	38.33333	6—Pre-Motor and Supplementary Motor Cortex	0.47826
20	−57.3333	−29.3333	53.66667	2—Primary Somatosensory Cortex	0.42748
21	−69	−30.6667	25.66667	40—Supramarginal gyrus part of Wernicke's area	0.68012
22	−64	−44.3333	42.33333	40—Supramarginal gyrus part of Wernicke's area	1

Appendix D: Full List of the Experimental Stimuli of Chap. 7

Category	No.	Prime			Target		
		Stimuli	Pinyin	Meaning	Stimuli	Pinyin	Meaning
Derivation	1	空	kōng	air	空姐	kōng jiě	stewardess
Derivation	2	贴	tiē	post	贴吧	tiē bā	post bar
Derivation	3	叔	shū	uncle	叔控	shū kòng	uncle fan
Derivation	4	型	xíng	stylish	型男	xíng nán	stylish man
Derivation	5	体	tǐ	sports	体坛	tǐ tán	sports circle
Derivation	6	学	xué	student	学潮	xué cháo	student unrest
Derivation	7	媒	méi	match	媒人	méi rén	matchmaker
Derivation	8	局	jú	bureau	局长	jú zhǎng	director general
Derivation	9	韩	hán	Korean	韩式	hán shì	Korean style
Derivation	10	笑	xiào	laugh	笑星	xiào xīng	comedian
Derivation	11	猎	liè	hunt	猎户	liè hù	hunter
Derivation	12	网	wǎng	net	网民	wǎng mín	netizen
Derivation	13	法	fǎ	law	法盲	fǎ máng	legal illiterate
Derivation	14	蚁	yǐ	ant	蚁族	yǐ zú	ant tribe
Derivation	15	书	shū	book	书虫	shū chóng	bookworm
Derivation	16	宅	zhái	otaku	宅女	zhái nǚ	otaku girl
Derivation	17	学	xué	study	学界	xué jiè	academia
Derivation	18	手	shǒu	hand	手感	shǒu gǎn	hand feeling
Derivation	19	国	guó	nation	国别	guó bié	nationality
Derivation	20	军	jūn	military	军嫂	jūn sǎo	military spouse
Derivation	21	海	hǎi	ocean	海员	hǎi yuán	sailor
Derivation	22	拳	quán	boxing	拳手	quán shǒu	boxer
Derivation	23	车	chē	cart	车夫	chē fū	carter
Derivation	24	恶	è	evil	恶棍	è gùn	vallain
Derivation	25	巫	wū	witchcraft	巫师	wū shī	witcher
Derivation	26	枣	zǎo	date	枣泥	zǎo ní	date paste
Derivation	27	药	yào	medicine	药学	yào xué	pharmacy

F. Gao, *The When, Where, and How of Morphological Processing in the Human Brain*, https://doi.org/10.1007/978-981-95-5981-7

Category	No.	Prime			Target		
		Stimuli	Pinyin	Meaning	Stimuli	Pinyin	Meaning
Derivation	28	贡	gòng	pay tribute	贡品	gòng pǐn	tribute
Derivation	29	球	qiú	ball game	球迷	qiú mí	ball game fan
Derivation	30	武	wǔ	martial	武士	wǔ shì	warrior
Derivation	31	考	kǎo	examine	考生	kǎo shēng	examinee
Derivation	32	农	nóng	agricultural	农业	nóng yè	agriculture
Derivation	33	花	huā	flower	花儿	huā ér	flower
Derivation	34	暗	àn	dark	暗处	àn chù	dark place
Derivation	35	咱	zán	I	咱们	zán men	we
Derivation	36	赌	dǔ	gamble	赌鬼	dǔ guǐ	gambler
Derivation	37	毒	dú	toxic	毒性	dú xìng	toxicity
Derivation	38	画	huà	paint	画家	huà jiā	painter
Derivation	39	石	shí	stone	石头	shí tou	stone
Derivation	40	桌	zhuō	table	桌子	zhuō zi	table
Compound	1	性	xìng	quality	性情	xìng qíng	temperament
Compound	2	音	yīn	voice	音讯	yīn xùn	message
Compound	3	江	jiāng	river	江湖	jiāng hú	rivers and lakes
Compound	4	道	dào	morality	道义	dào yì	morality and justice
Compound	5	口	kǒu	mouth	口吻	kǒu wěn	tone of voice
Compound	6	声	shēng	sound	声势	shēng shì	momentum
Compound	7	手	shǒu	hand	手足	shǒu zú	kinship
Compound	8	周	zhōu	periphery	周围	zhōu wéi	around
Compound	9	朝	cháo	dynasty	朝代	cháo dài	dynasty
Compound	10	传	zhuàn	biography	传记	zhuàn jì	biography
Compound	11	群	qún	crowd	群众	qún zhòng	the masses
Compound	12	姿	zī	position	姿势	zī shì	position
Compound	13	功	gōng	achievement	功名	gōng míng	fame
Compound	14	婆	pó	mother-in-law	婆媳	pó xí	mother-in-law and daughter-in-law
Compound	15	堕	duò	fall	堕落	duò luò	corrupt
Compound	16	胳	gē	arm	胳膊	gē bo	arm
Compound	17	符	fú	sign	符号	fú hào	symbol
Compound	18	沐	mù	wash	沐浴	mù yù	shower
Compound	19	书	shū	letter	书信	shū xìn	letter
Compound	20	岁	suì	year	岁月	suì yuè	years
Compound	21	缝	fèng	crack	缝隙	fèng xì	gap
Compound	22	诅	zǔ	swear	诅咒	zǔ zhòu	curse
Compound	23	宇	yǔ	space	宇宙	yǔ zhòu	universe
Compound	24	朝	zhāo	daytime	朝夕	zhāo xī	day and night
Compound	25	笔	bǐ	pen	笔墨	bǐ mò	writing
Compound	26	沼	zhǎo	marsh	沼泽	zhǎo zé	marsh
Compound	27	风	fēng	wind	风雨	fēng yǔ	wind and rain

Category	No.	Prime			Target		
		Stimuli	Pinyin	Meaning	Stimuli	Pinyin	Meaning
Compound	28	路	lù	way	路径	lù jìng	path
Compound	29	脂	zhī	fat	脂肪	zhī fáng	fat
Compound	30	锅	guō	pot	锅炉	guō lú	boiler
Compound	31	花	huā	flower	花草	huā cǎo	plant
Compound	32	篇	piān	section	篇章	piān zhāng	writings
Compound	33	皮	pí	skin	皮肤	pí fū	skin
Compound	34	田	tián	field	田园	tián yuán	countryside
Compound	35	班	bān	class	班级	bān jí	class
Compound	36	翅	chì	wing	翅膀	chì bǎng	wing
Compound	37	财	cái	estate	财宝	cái bǎo	treasure
Compound	38	村	cūn	village	村庄	cūn zhuāng	village
Compound	39	街	jiē	street	街道	jiē dào	street
Compound	40	针	zhēn	needle	针线	zhēn xiàn	needlework
Non-morphological	1	脏	zāng	dirty	污水	wū shuǐ	waste water
Non-morphological	2	乐	lè	happy	微笑	wēi xiào	smile
Non-morphological	3	饭	fàn	meal	早餐	zǎo cān	breakfast
Non-morphological	4	庙	miào	temple	佛教	fó jiào	Buddhism
Non-morphological	5	写	xiě	write	纸张	zhǐ zhāng	paper
Non-morphological	6	醉	zuì	drunk	白酒	bái jiǔ	liquor
Non-morphological	7	狱	yù	prison	铁窗	tiě chuāng	prison
Non-morphological	8	哭	kū	cry	悲剧	bēi jù	tragedy
Non-morphological	9	爬	pá	climb	登山	dēng shān	mountaineering
Non-morphological	10	学	xué	study	上课	shàng kè	go to class
Non-morphological	11	酸	suān	sour	陈醋	chén cù	mature vinegar
Non-morphological	12	晚	wǎn	night	半夜	bàn yè	midnight
Non-morphological	13	火	huǒ	fire	燃烧	rán shāo	burn
Non-morphological	14	吵	chǎo	noisy	噪声	zào shēng	noise
Non-morphological	15	穿	chuān	wear	服装	fú zhuāng	clothing
Non-morphological	16	树	shù	tree	森林	sēn lín	forest
Non-morphological	17	晴	qíng	sunny	蓝天	lán tiān	blue sky
Non-morphological	18	考	kǎo	examine	试卷	shì juàn	examination paper
Non-morphological	19	想	xiǎng	think	沉思	chén sī	ruminate
Non-morphological	20	皇	huáng	royal	国王	guó wáng	king
Non-morphological	21	热	rè	hot	烈日	liè rì	burning sun
Non-morphological	22	道	dào	way	马路	mǎ lù	road
Non-morphological	23	闷	mēn	bored	无聊	wú liáo	bored
Non-morphological	24	老	lǎo	old	高龄	gāo líng	old age
Non-morphological	25	卖	mài	sell	促销	cù xiāo	on sale
Non-morphological	26	走	zǒu	walk	远足	yuǎn zú	hiking
Non-morphological	27	房	fáng	house	住宅	zhù zhái	residence
Non-morphological	28	病	bìng	sickness	医院	yī yuàn	hospital

Category	No.	Prime			Target		
		Stimuli	Pinyin	Meaning	Stimuli	Pinyin	Meaning
Non-morphological	29	睡	shuì	sleep	休息	xiū xi	rest
Non-morphological	30	神	shén	deity	上帝	shàng dì	God
Non-morphological	31	内	nèi	inner	里边	lǐ bian	inside
Non-morphological	32	买	mǎi	buy	采购	cǎi gòu	purchase
Non-morphological	33	送	sòng	give	馈赠	kuì zèng	present a gift
Non-morphological	34	因	yīn	cause	缘故	yuán gù	reason
Non-morphological	35	智	zhì	intelligence	聪明	cōng míng	smart
Non-morphological	36	聋	lóng	deaf	失聪	shī cōng	deafness
Non-morphological	37	累	lèi	tired	疲惫	pí bèi	exhausted
Non-morphological	38	兄	xiōng	elder brother	哥哥	gē ge	elder brother
Non-morphological	39	赞	zàn	like	表扬	biǎo yáng	praise
Non-morphological	40	丑	chǒu	hideous	难看	nán kàn	ugly

Appendix E: Full List of the Chinese Stimuli of Chap. 8

Priming type	Prime			Target			People-related
	Stimuli	Pinyin	Meaning	Stimuli	Pinyin	Meaning	
Morphological	书法	shū fǎ	calligraphy	书法家	shū fǎ jiā	calligrapher	Yes
Morphological	小气	xiǎo qì	stingy	小气鬼	xiǎo qì guǐ	miser	Yes
Morphological	服务	fú wù	service	服务员	fú wù yuán	waiter	Yes
Morphological	建筑	jiàn zhù	architecture	建筑师	jiàn zhù shī	architect	Yes
Morphological	拳击	quán jī	boxing	拳击手	quán jī shǒu	boxer	Yes
Morphological	骑	qí	ride	骑士	qí shì	knight	Yes
Morphological	研究	yán jiū	research	研究生	yán jiū shēng	graduate	Yes
Morphological	受害	shòu hài	fall victim	受害人	shòu hài rén	victim	Yes
Morphological	追星	zhuī xīng	starchasing	追星族	zhuī xīng zú	idolater	Yes
Morphological	车	chē	cart	车夫	chē fū	carter	Yes
Morphological	暴发	bào fā	upstart	暴发户	bào fā hù	overnight millionaire	Yes
Morphological	委员	wěi yuán	committee	委员长	wěi yuán zhǎng	chairman	Yes
Morphological	歌	gē	song	歌手	gē shǒu	singer	Yes
Morphological	笑	xiào	laugh	笑星	xiào xīng	comedian	Yes
Morphological	渔	yú	fishing	渔民	yú mín	fisherman	Yes
Morphological	音乐	yīn yuè	music	音乐家	yīn yuè jiā	musician	Yes
Morphological	理发	lǐ fà	haircut	理发师	lǐ fà shī	barber	Yes
Morphological	舞蹈	wǔ dǎo	dance	舞蹈家	wǔ dǎo jiā	dancer	Yes
Morphological	基督	jī dū	Christianity	基督徒	jī dū tú	Christian	Yes
Morphological	男	nán	male	男儿	nán ér	man	Yes
Morphological	球	qiú	ball game	球迷	qiú mí	ball game fan	Yes
Morphological	恶	è	evil	恶棍	è gùn	vallain	Yes

F. Gao, *The When, Where, and How of Morphological Processing in the Human Brain*, https://doi.org/10.1007/978-981-95-5981-7

Priming type	Prime			Target			People-related
	Stimuli	Pinyin	Meaning	Stimuli	Pinyin	Meaning	
Morphological	工作	gōng zuò	work	工作狂	gōng zuò kuáng	workaholic	Yes
Morphological	妈妈	mā ma	mother	准妈妈	zhǔn mā ma	mother-to-be	Yes
Morphological	留学	liú xué	study abroad	留学生	liú xué shēng	overseas student	Yes
Morphological	化妆	huà zhuāng	make-up	化妆师	huà zhuāng shī	dresser	Yes
Morphological	飞行	fēi xíng	flight	飞行员	fēi xíng yuán	pilot	Yes
Morphological	电	diàn	electricity	电工	diàn gōng	electrician	Yes
Morphological	流浪	liú làng	vagrancy	流浪者	liú làng zhě	vagrant	Yes
Morphological	死	sǐ	dead	死者	sǐ zhě	the decreased	Yes
Morphological	音乐	yīn yuè	music	音乐界	yīn yuè jiè	the music industry	No
Morphological	世界	shì jiè	world	世界观	shì jiè guān	world view	No
Morphological	进化	jìn huà	evolution	进化论	jìn huà lùn	theory of evolution	No
Morphological	人类	rén lèi	human	人类学	rén lèi xué	anthropology	No
Morphological	高	gāo	high	高度	gāo dù	altitude	No
Morphological	方向	fāng xiàng	direction	方向感	fāng xiàng gǎn	sense of direction	No
Morphological	便携	biàn xié	portable	便携式	biàn xié shì	portable	No
Morphological	留学	liú xué	studying abroad	留学热	liú xué rè	craze in studying abroad	No
Morphological	可行	kě xíng	feasible	可行性	kě xíng xìng	feasibility	No
Morphological	现代	xiàn dài	modern	现代化	xiàn dài huà	modernization	No
Morphological	具体	jù tǐ	specify	具体化	jù tǐ huà	specification	No
Morphological	体	tǐ	body	体坛	tǐ tán	sports circle	No
Morphological	游泳	yóu yǒng	swimming	游泳馆	yóu yǒng guǎn	natatorium	No
Morphological	歌舞	gē wǔ	singing and dancing	歌舞界	gē wǔ jiè	singing and dancing circle	No
Morphological	房	fáng	house	房子	fáng zi	house	No
Morphological	正式	zhèng shì	formal	非正式	fēi zhèng shì	informal	No
Morphological	动物	dòng wù	animal	动物园	dòng wù yuán	zoo	No
Morphological	自动	zì dòng	automatic	半自动	bàn zì dòng	semi-automatic	No
Morphological	健康	jiàn kāng	health	亚健康	yà jiàn kāng	sub-health	No
Morphological	化妆	huà zhuāng	make-up	化妆品	huà zhuāng pǐn	cosmetics	No

Priming type	Prime			Target			People-related
	Stimuli	Pinyin	Meaning	Stimuli	Pinyin	Meaning	
Morphological	木	mù	wood	木头	mù tou	wood	No
Morphological	电子	diàn zǐ	electronic	电子版	diàn zǐ bǎn	electronic edition	No
Morphological	蒜	suàn	garlic	蒜泥	suàn ní	mashed garlic	No
Morphological	责任	zé rèn	responsibility	责任感	zé rèn gǎn	sense of responsibility	No
Morphological	现实	xiàn shí	realistic	超现实	chāo xiàn shí	surreal	No
Morphological	思	sī	ideology	思潮	sī cháo	ideological trend	No
Morphological	计算	jì suàn	computing	云计算	yún jì suàn	cloud computing	No
Morphological	电	diàn	electricity	电费	diàn fèi	electricity bill	No
Morphological	喷气	pēn qì	jet	喷气式	pēn qì shì	jet-propelled	No
Morphological	花	huā	flower	花儿	huā ér	flower	No
Semantic	大海	dà hǎi	sea	海盗	hǎi dào	pirate	Yes
Semantic	老师	lǎo shī	teacher	先生	xiān shēng	teacher	Yes
Semantic	友谊	yǒu yì	friendship	朋友	péng yǒu	friend	Yes
Semantic	妈	mā	mother	母亲	mǔ qīn	mother	Yes
Semantic	女孩	nǚ hái	girl	少女	shào nǚ	girl	Yes
Semantic	小偷	xiǎo tōu	thief	警察	jǐng chá	policeman	Yes
Semantic	公司	gōng sī	company	总裁	zǒng cái	president	Yes
Semantic	战争	zhàn zhēng	war	英雄	yīng xióng	hero	Yes
Semantic	家	jiā	home	夫妇	fū fù	couple	Yes
Semantic	监狱	jiān yù	prison	罪犯	zuì fàn	criminal	Yes
Semantic	保安	bǎo ān	safeguard	保镖	bǎo biāo	bodyguard	Yes
Semantic	官员	guān yuán	officer	书记	shū jì	secretary	Yes
Semantic	崇拜	chóng bài	worship	偶像	ǒu xiàng	idol	Yes
Semantic	退休	tuì xiū	retirement	老干部	lǎo gàn bù	veteran cadre	Yes
Semantic	收租	shōu zū	collect rents	房东	fáng dōng	landlord	Yes
Semantic	婚	hūn	marriage	未婚妻	wèi hūn qī	fiancee	Yes
Semantic	公车	gōng chē	bus	乘客	chéng kè	passenger	Yes
Semantic	叔	shū	uncle	阿姨	ā yí	aunt	Yes
Semantic	白宫	bái gōng	the White House	副总统	fù zǒng tǒng	vice president	Yes
Semantic	公主	gōng zhǔ	princess	皇后	huáng hòu	queen	Yes
Semantic	权力	quán lì	power	主席	zhǔ xí	chairman	Yes
Semantic	经理	jīng lǐ	manager	主管	zhǔ guǎn	supervisor	Yes
Semantic	主人	zhǔ rén	master	奴隶	nú lì	slave	Yes

Priming type	Prime			Target			People-related
	Stimuli	Pinyin	Meaning	Stimuli	Pinyin	Meaning	
Semantic	外公	wài gōng	grandfather	外婆	wài pó	grandmother	Yes
Semantic	儒	rú	Confucian	孔子	kǒng zǐ	Confucius	Yes
Semantic	信任	xìn rèn	trust	同伴	tóng bàn	partner	Yes
Semantic	时尚	shí shàng	fashion	模特	mó tè	model	Yes
Semantic	姊妹	zǐ mèi	sister	三姐妹	sān jiě mèi	three sisters	Yes
Semantic	孪生	luán shēng	twinborn	双胞胎	shuāng bāo tāi	twins	Yes
Semantic	班级	bān jí	class	班主任	bān zhǔ rèn	head teacher	Yes
Semantic	晴	qíng	sunny	蓝天	lán tiān	blue sky	No
Semantic	答案	dá àn	answer	问题	wèn tí	question	No
Semantic	钟	zhōng	clock	时间	shí jiān	time	No
Semantic	现在	xiàn zài	now	今天	jīn tiān	today	No
Semantic	游戏	yóu xì	game	比赛	bǐ sài	competition	No
Semantic	企业	qǐ yè	enterprise	公司	gōng sī	company	No
Semantic	自拍	zì pāi	selfie	照片	zhào piàn	photo	No
Semantic	钱	qián	money	人民币	rén mín bì	RMB	No
Semantic	表演	biǎo yǎn	performance	晚会	wǎn huì	gala	No
Semantic	茶	chá	tea	咖啡	kā fēi	coffee	No
Semantic	睡	shuì	sleep	休息	xiū xī	rest	No
Semantic	病	bìng	sickness	医院	yī yuàn	hospital	No
Semantic	法律	fǎ lǜ	law	秩序	zhì xù	order	No
Semantic	住所	zhù suǒ	residence	公寓	gōng yù	apartment	No
Semantic	影院	yǐng yuàn	theater	电影	diàn yǐng	movie	No
Semantic	酒吧	jiǔ bā	bar	俱乐部	jù lè bù	club	No
Semantic	女神	nǚ shén	goddess	希腊	xī là	Greece	No
Semantic	神经	shén jīng	nerve	脑	nǎo	brain	No
Semantic	吃	chī	eat	美食	měi shí	delicacy	No
Semantic	政府	zhèng fǔ	government	中南海	zhōng nán hǎi	Zhongnanhai	No
Semantic	旅游	lǚ yóu	tourism	酒店	jiǔ diàn	hotel	No
Semantic	和尚	hé shàng	monk	寺庙	sì miào	temple	No
Semantic	明星	míng xīng	celebrity	娱乐	yú lè	entertainment	No
Semantic	欧洲	ōu zhōu	Europe	西班牙	xī bān yá	Spain	No
Semantic	电话	diàn huà	telephone	手机	shǒu jī	cellphone	No
Semantic	耳	ěr	ear	听觉	tīng jué	auditory sense	No

	Prime			Target			People-related
Priming type	Stimuli	Pinyin	Meaning	Stimuli	Pinyin	Meaning	
Semantic	积极	jī jí	active	正能量	zhèng néng liàng	positive energy	No
Semantic	内蒙	nèi měng	Inner Mongolia	高原	gāo yuán	plateau	No
Semantic	发票	fā piào	invoice	报销	bào xiāo	reimbursement	No
Semantic	法官	fǎ guān	judge	正义	zhèng yì	justice	No

Appendix F: Full List of the English Stimuli of Chap. 8

Morphological Priming				Semantic Priming			
People-related		Non people-related		People-related		Non people-related	
prime	target	prime	target	prime	target	prime	target
teach	teacher	happy	happiness	fly	bird	work	partner
music	musician	able	ability	road	way	hospital	nurse
farm	farmer	act	action	drive	car	double	twins
visit	visitor	slow	slowness	give	gift	woman	lady
educate	educator	create	creation	music	piano	royal	king
dream	dreamer	educate	education	gold	silver	old	grandma
library	librarian	move	movement	university	college	wife	husband
drive	driver	state	statement	wind	rain	uncle	aunt
wait	waitress	kind	kindness	nature	science	young	son
sail	sailor	suggest	suggestion	cream	milk	England	queen
city	citizen	simple	simplicity	past	future	company	boss
science	scientist	difficult	difficulty	train	subway	sales	clerk
history	historian	invent	invention	doctor	medicine	home	housewife
million	millionaire	direct	directness	photo	picture	leader	president
foreign	foreigner	most	mostly	sand	beach	palace	prince
run	runner	sun	sunny	language	speech	master	expert
technique	technician	dry	dryer	test	exam	sister	brother
act	actress	fear	fearless	science	biology	school	student
politics	politician	peace	peaceful	sports	basketball	lady	gentleman
organize	organizer	locate	location	club	bar	father	daughter
sing	singer	question	questionnaire	years	age	kid	children
art	artist	store	storage	city	province	guy	girl
serve	servant	arrive	arrival	factory	product	person	people
invent	inventor	accurate	accuracy	rose	flower	house	parents
direct	director	serve	service	week	Friday	uncle	cousin
strange	stranger	friend	friendship	egg	breakfast	son	daddy

F. Gao, *The When, Where, and How of Morphological Processing in the Human Brain*, https://doi.org/10.1007/978-981-95-5981-7

Morphological Priming				Semantic Priming			
People-related		Non people-related		People-related		Non people-related	
employ	employee	feel	feeling	story	book	classroom	classmate
interview	interviewee	decide	decision	carry	bag	welcome	guest
assist	assistant	possible	possibility	nurse	hospital	buddy	friend
magic	magician	safe	safety	name	address	mother	baby

MIX
Papier aus verantwortungsvollen Quellen
Paper from responsible sources
FSC® C105338

If you have any concerns about our products,
you can contact us on
ProductSafety@springernature.com

In case Publisher is established outside the EU,
the EU authorized representative is:
Springer Nature Customer Service Center GmbH
Europaplatz 3, 69115 Heidelberg, Germany

Printed by Libri Plureos GmbH
in Hamburg, Germany